AF802801

[redacted]

FLUID, ELECTROLYTE, AND ACID-BASE REGULATION

[redacted]

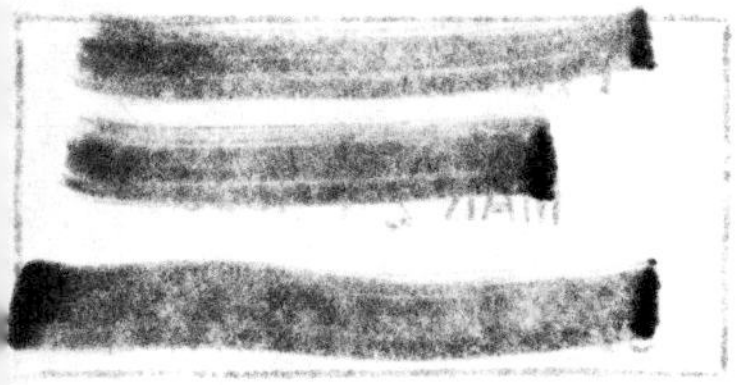

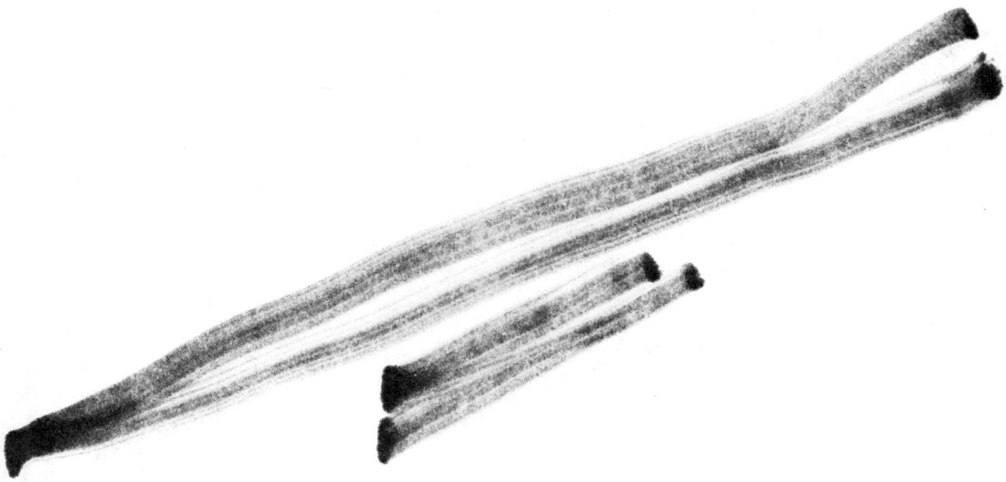

In print $25
$5
3.50

FLUID, ELECTROLYTE, AND ACID-BASE REGULATION

Jack L. Keyes
Linfield-Good Samaritan School of Nursing

Wadsworth Health Sciences Division
Monterey, California

Wadsworth Health Sciences Division
A Division of Wadsworth, Inc.

© 1985 by Wadsworth, Inc., Belmont, California 94002. All rights reserved. No part of this book may be reproduced, stored in a retrieval system, or transcribed, in any form or by any means—electronic, mechanical, photocopying, recording, or otherwise—without the prior written permission of the publisher, Wadsworth Health Sciences Division, Monterey, California 93940, a division of Wadsworth, Inc.

Printed in the United States of America
10 9 8 7 6 5 4 3 2 1

Library of Congress Cataloging in Publication Data

Keyes, Jack L., [date]
Fluid, electrolyte, and acid-base regulation.

Bibliography: p.
Includes index.
1. Body fluid disorders. 2. Water-electrolyte imbalances. 3. Acid-base imbalances. 4. Homeostasis.
I. Title.
RC630.K49 1985 616.3'9 84-21925
ISBN 0-534-04422-0

Sponsoring Editor: *Thomas J. Keating*
Editorial Assistants: *Mary Beth McDavid and Corinne Kibbe*
Production Editor: *Constance D. Brown*
Manuscript Editor: *Dorothy Hoffman*
Permissions Editor: *Mary Kay Hancharick*
Interior Design: *Brenn Lea Pearson*
Cover Design: *Sharon Smith*
Cover Illustration: © *Stanley Rice*
Art Coordinator: *Judy Macdonald*
Interior Illustration: *Cyndie Clark-Huegel*
Typesetting: *Science Typographers, Inc., Medford, Long Island, New York*
Printing and Binding: *R. R. Donnelley & Sons Co., Crawfordsville, Indiana*

To my wife, Connie, and my children, Jim and Tanya,
who encouraged me and gave up so much
that I might complete this endeavor.

PREFACE

This is an intermediate to advanced text on the physiology and pathophysiology of fluid, electrolyte, and acid-base regulation. It is intended for students and health care professionals who are engaged in caring for patients with disturbances of fluid, electrolyte, and acid-base balance in any of the myriad of clinical settings. I have attempted to provide in a single source both normal physiology and pathophysiology of this complex topic. The approach used in this book has been tested on students for over 15 years. Techniques and approaches that did not work have been discarded and those that were successful have been polished and are presented here.

Several features of this book facilitate learning. The topics presented build from the simple to the more complex and from normal function to disordered function. Application to the clinical setting is frequently the most difficult, and a baker's dozen of case examples are presented in Chapter 10. The topic of transport, which is an essential base for all of this content, appears in the first chapter. In addition, there are full separate chapters on blood-gas transport and hypoxia. An extensive discussion of buffering is presented in a separate chapter because it is the basis for understanding both normal and disordered acid–base regulation. Compensation is presented as a set of processes separate and distinct from buffering. There is also a section on venous blood-gas composition, which is becoming more important in this field. The emphasis of the book is on principles and regulation. This approach discourages rote memory and enhances learning that can be applied to new situations.

The chapters of the book are intended to be read in sequence, however, other organizational approaches are also useful. For example, once Chapter 1 has been learned, the topics may be read as blocks of material. Chapters 2, 7, and 8 comprise a block on the subject of fluid and electrolyte balance. Chapters 3, 4, and 9 are a block on the topic of blood gases and Chapters 4, 5, and 6 may be read as a separate section on acid-base regulation. Appendices provide additional information for more advanced topics such as Donnan equilibrium, anion gap, and base excess and deficit. Where appropriate, derivations of equations are presented. The reader who has no interest in those can skip the derivations and proceed to the final form of the equation. The derivations do provide the interested learner with the assumptions and the reason for the final form of the equation.

A book of this type does not lend itself to extensive references and citations. To offset this deficit, I have included at the end of the book a list of several references and readings that will provide more depth and detailed references for each chapter.

ACKNOWLEDGMENTS

I would like to express my deep appreciation to those who helped make this book possible. First, to John M. Brookhart, Ph.D., and Robert E. Swanson, Ph.D., outstanding teachers and mentors. Their disciplined scholarship and uncompromising integrity have been an inspiration and example to me. It is my great and good fortune to call them friends. Special thanks are extended to Kent Thornburg, Ph.D., and Dan Stiffler, Ph.D., who read the rough draft of this manuscript and provided constructive criticism when it was most sorely needed. Many students, too numerous to list, read parts of this manuscript and also provided helpful comments. The compensation curves shown in Figure 6-4 and the approach used for osmosis were developed by R. E. Swanson, Ph.D., and are used here with his kind permission. Finally, appreciation is extended to Diane Dean, Carol Johnson, and Connie Keyes for helping type the manuscript.

Jack L. Keyes

CONTENTS

Fundamental Concepts Central to Fluid-Electrolyte and Acid-Base Regulation

CHAPTER 1

Body cells live in a special environment called interstitial fluid (or ISF). This environment has a unique composition of electrolyte concentrations and pH that is essential for maintaining normal function. For example, potassium concentration in the ISF is maintained within the narrow limits of 3.5 to 5.0 milliequivalents per liter (mEq/L). This narrow range is necessary for maintaining electrical activity of excitable cells such as those of the cardiac conduction system and nervous system. The hydrogen ion concentration, $[H^+]$, is regulated within even narrower limits of 37 to 43 nanoequivalents per liter (nEq/L). The $[H^+]$ of ISF is, thus, slightly more alkaline than that of pure water, which has an $[H^+]$ of 100 nEq/L (Figure 1-1). The alkaline pH of ISF is essential for normal function of enzymes located on cell surfaces. Even the concentrations of gases such as oxygen and carbon dioxide are regulated and must be maintained within a specific range if cell activity is to remain optimum.

Because the concentrations of electrolytes and gases in ISF affect cell function, the body needs to be able to compensate for disturbances in volume and composition of both ISF and plasma. In disease or following trauma, however, fluid balance, electrolyte composition, and acid-base regulation may all be abnormal. Frequently, it is necessary to help the body adjust for these disturbances in balance. The following case illustrates this point.

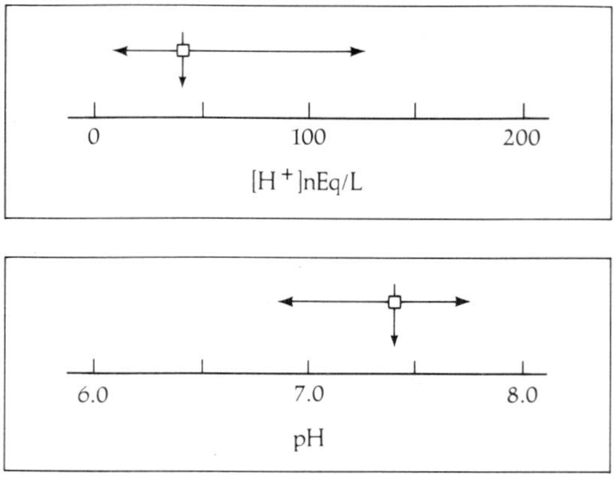

FIGURE 1-1
Normal range and values for hydrogen ion concentration, $[H^+]$, (*upper figure*) and pH (*lower figure*). The horizontal arrows in both scales show the range compatible with mammalian life. These are 16 to 126 nEq / L for $[H^+]$ and 6.9 to 7.8 for pH. The normal range is shown by the horizontal limits of the squares whereas the normal values for each parameter are shown by the vertical line. Normal values for $[H^+]$ and pH in arterial plasma are 40 nEq / L and 7.4, respectively.

Case 1 A 54-year-old man was admitted to a large metropolitan hospital for surgical repair of an abdominal aortic aneurysm. After surgery, the patient was brought to the surgical intensive care unit. An endotracheal tube, nasogastric tube, and urinary catheter were in place. He was mechanically ventilated with an inspired gas mixture containing 40% O_2 and 60% N_2. His tidal volume was set at 950 ml and he was breathing at a frequency of 12 breaths per minute. He was receiving a diuretic (Lasix®) twice a day and morphine for pain. Fluid therapy consisted of 2 L of 5% glucose in water (D5W) every 24 hours. Forty-eight hours after surgery the results of arterial blood-gas analysis were: pH = 7.62, P_{CO_2} = 26 torr, $[HCO_3]$ = 27 mEq/L, and P_{O_2} = 74 torr. Serum electrolyte concentrations were $[Na^+]$ = 126 mEq/L, $[K^+]$ = 3.4 mEq/L, $[Cl^-]$ = 83 mEq/L, and $[Ca^{++}]$ = 4.8 mEq/L.

This man had both fluid-electrolyte and acid-base disturbances. The following questions may be answered from the data presented in the case: What kind of acid-base disturbance is present? What is the cause of the electrolyte disturbance? What are the causes of the acid-base imbalance? What, if anything, should be done to intervene? To answer these questions it is first necessary to define normal acid-base status as well as the normal processes of fluid and electrolyte balance. Second, it is also necessary to determine the processes taking place during normal as well as disordered acid-base status and finally to interpret data obtained from both serum electrolyte and blood-gas analyses. After mastering the material in this text, the reader will be able to answer the questions asked about Case 1.

It should be noted, before proceeding further into this text, that acid-base regulation is a part of fluid and electrolyte regulation. These subjects are easier to learn if acid-base regulation is presented somewhat separately from fluid and electrolyte regulation. As a result, several chapters deal primarily with acid-base physiology and others deal primarily with fluid and electrolyte regulation. The reader must remember, however, that when disturbances of fluid and electrolyte balance occur, disturbances of acid-base balance will frequently occur as well.

FUNDAMENTAL CONCEPTS AND DEFINITIONS

In the chapters that follow, new terms are defined when they are presented for the first time. However, the terms and concepts discussed in this section are fundamental prerequisites to understanding material presented throughout the text. Many of the terms relate directly to the definitions of acids and bases. The definitions for acids and bases devised by Bronstedt and Lowry (1923) are used in this text because they provide the best framework on which to build knowledge of fluid, electrolyte, and acid-base physiology. Mastery of the terms and concepts presented in this section will provide insight and understanding of the more complex concepts that follow.

Electrolyte An *electrolyte* is a chemical compound that forms ions in solution. Solutions of electrolytes are capable of conducting an electric current. An *ion* is an atom or molecule that is electrically charged. The charge on the ion results when the atom or molecule gains or loses one or more electrons. When electrons are gained, the atoms become negatively charged and are called *anions*. When electrons are lost, the atoms become positively charged and are called *cations*.

Some electrolytes are always ionized, even when they are not in solution. For example, table salt (or sodium chloride) is ionized in the dry crystalline state. In body fluids all of the sodium and chloride ions are separated from each other. In other words, they are not *associated* (joined together) but rather *dissociated*. Electrolytes that are completely dissociated in solution are said to be *strong electrolytes* whereas those that are not completely dissociated are said to be *weak electrolytes*. Electrolytes in body fluids that are composed of sodium (Na^+) or potassium (K^+) ions are strong electrolytes.

Acids and Bases An *acid* is defined as a hydrogen ion donor. A hydrogen ion is the proton nucleus of a hydrogen atom and is a positively charged ion (i.e., a cation). Any substance that ionizes (dissociates) to form an H^+ in solution is an acid (Equation 1-1). The substance HB in Reaction 1-1 is an acid. Hydrogen ions are very reactive species.

$$HB \rightarrow H^+ + B^- \tag{1-1}$$

For all practical purposes an aqueous solution contains no free H^+. These protons, when formed, combine rapidly with neutral water molecules to form the hydronium ion, H_3O^+ (Equation 1-2).

$$H^+ + H_2O \rightleftharpoons H_3O^+ \tag{1-2}$$

The hydronium ion gives a solution its acidity or acid character. However, the more conventional term, hydrogen ion, and its chemical symbol, H^+, will be used in this text in lieu of hydronium ion. A *base* is an acceptor of hydrogen ions. In solution, bases combine chemically (associate) with hydrogen ions. A base neutralizes the effects of H^+ by combining with the ions and thereby removing them from activity in the solution.

$$B^- + H^+ \rightarrow HB \tag{1-3}$$

When an acid dissociates, it produces a hydrogen ion and another specie (ion or neutral molecule) called a *conjugate* base. In Reaction 1-1 above the conjugate base is B^-. When a base associates with a hydrogen ion, it forms another ion or neutral molecule called a *conjugate acid*. In Reaction 1-3 the conjugate acid is HB.

Equilibrium Equilibrium is a special kind of steady state in which the rate of a reaction in one direction is equal to the rate of the reverse reaction. If the reactions in Equations 1-1 and 1-3 above are combined,

$$HB \rightleftharpoons H^+ + B^- \tag{1-4}$$

at equilibrium the rate of dissociation would equal the rate of association. The concentrations of the various components do not have to be equal; only the rates of dissociation and association have to be equal at equilibrium. In Reaction 1-4 the rate of dissociation is directly proportional to the concentration of HB. The rate of association is directly proportional to the product of the concentrations of H^+ and B^-. In mathematical terms, the rates (or velocities) of dissociation and association may be summarized as follows:

$$V_d = k_1[HB] \tag{1-5}$$

where V_d is the rate of dissociation and k_1 is a proportionality constant, and

$$V_a = k_2[H^+][B^-] \tag{1-6}$$

where V_a is the rate of association and k_2 is a proportionality constant.

From Equations 1-5 and 1-6, it can be seen that if the concentrations of HB, H^+, or B^- are altered, velocities of the association and dissociation reactions will also be altered. For example, increasing the concentration of HB will increase the rate of dissociation, whereas decreasing the concentration of HB has the opposite effect. A similar argument can be made for changing the rate of association when either $[H^+]$ or $[B^-]$ is increased or decreased. Therefore, when the concentration of any component of the reaction in Equation 1-4 is changed, the velocities of the individual reactions will change until equilibrium is again established. For example, if $[B^-]$ is increased, there will be an increase in the rate of association. This increased association increases [HB] and decreases $[H^+]$. The increased [HB], in turn, causes an increase in V_d. The resulting decreased $[H^+]$ reduces V_a. As a result, V_a and V_d will again become equal, and a new equilibrium will be established with increased [HB], increased

Table 1-1 Conjugate Acid-Base Pairs

Name*	Conjugate acid	Conjugate base	Acidity	Ka	pKa
Hydrochloric acid	HCl	Cl^-	strong		
Sulfuric acid	H_2SO_4	HSO_4^-	strong		
Lactic acid	HLac	Lac^-	weak	1.4×10^{-4}	3.85
Acetoacetic acid†	HAcet	$Acet^-$	weak	2.63×10^{-4}	3.58
β-Hydroxybutyric acid†	HβHyd	βHyd^-	weak	2.0×10^{-5}	4.70
Carbonic acid	H_2CO_3	HCO_3^-	weak	4.27×10^{-8}	6.37
Phosphoric acid	$H_2PO_4^-$	$HPO_4^=$	weak	6.2×10^{-8}	7.21
Ammonium ion	NH_4^+	NH_3	weak	5.6×10^{-10}	9.20
Hemoglobin	HHb	Hb^-	weak		
Protein	HPr	Pr^-	weak		

**All of these acids, with the possible exception of sulfuric acid, are found in body fluids. Values for Ka and pKa are listed for those weak acids that have a specific dissociation constant. The greater the value of Ka, the stronger the acid. Since pKa equals log 1/Ka, the values for pKa vary inversely to Ka and, therefore, to the strength of the acid. An acid or a base may be a cation, an anion, or a neutral molecule. In body fluids most bases are either anions or neutral molecules. The concept of pKa is discussed further in Appendix I.*

†Acetoacetic acid and β-hydroxybutryic acid are ketone bodies.

$[B^-]$, and decreased $[H^+]$. Normally, reestablishing equilibrium following a change in the concentration of either reactant or products requires less than 1 second in time.

Strong versus Weak Acids The distinction made between strong and weak acids is based on the degree of dissociation of the acid. A *strong acid* dissociates essentially completely into hydrogen ions and its conjugate base when in solution (Figure 1-2). As a result, there are virtually no conjugate acid molecules of the strong acid remaining in the solution. Examples of strong acids include the mineral acids such as hydrochloric and sulfuric acid (Table 1-1).

Strong Acids

$$\mathrm{HB} \rightleftarrows \mathbf{H^+ + B^-}$$

Weak Acids

$$\mathrm{HB} \rightleftarrows \mathrm{H^+ + B^-}$$

FIGURE 1-2
The distinction between strong and weak acids. The sizes of the letters of the reaction components indicate their relative concentrations. Strong acids in solution are almost totally dissociated. In fact, strong acids, such as HCl, can dissociate so completely that virtually no acid remains in solution. The concentrations of H^+ and B^- relative to HB are high. In contrast, weak acids dissociate to a lesser degree so that a finite concentration of acid remains in solution. Hence, the concentrations of H^+ and B^- relative to HB are much lower than with strong acids.

In contrast to strong acids, a *weak acid* does not dissociate completely in a solution. At equilibrium there is a finite concentration of each component (Equation 1-4, Figure 1-2). The strength or acidity of a solution depends on the degree of dissociation of the acid. Some weak acids are stronger, that is, more completely dissociated than other weak acids. For example, lactic acid is a stronger weak acid than carbonic acid at the same concentration because lactic acid is more completely dissociated at equilibrium than carbonic acid (Table 1-1).

Volatile versus Nonvolatile Acids A *volatile acid* is defined as an acid that can be excreted from the body as a gas. Either the acid itself or a chemical product of the acid can be converted to a gas and excreted. Carbonic acid dehydrates to carbon dioxide (CO_2) and water in the lungs where the CO_2 is excreted. Hence, carbonic acid is classified as a volatile acid.

Nonvolatile acids (or fixed acids) must be metabolized or excreted from the body in an aqueous solution. All acids found in body fluids other than carbonic acid are classified as nonvolatile acids. In many cases, it is not necessarily the acid itself that is excreted but the conjugate base of the acid that appears in the urine. Examples of fixed acids that are metabolized include lactic acid, acetoacetic acid, and β-hydroxybutyric acid (Table 1-1). Acids (or their conjugate bases) that are excreted in urine include HSO_4^-, NH_4^+, and $H_2PO_4^-$ (Table 1-1).

Concept of pH An expression of the acidity or alkalinity of a solution is *pH*. The acidity or alkalinity of a solution depends only on the hydrogen ion concentration (or $[H^+]$). Some clinicians favor assessing acid–base status in terms of $[H^+]$, but pH is equally useful and is more commonly used. The $[H^+]$ and pH are inversely related to each other, as shown in Equation 1-7

$$\text{pH} = \log \frac{1}{[H^+]} \qquad (1\text{-}7)$$

where $[H^+]$ is expressed in equivalents per liter. Note: pH does not refer to concentration of acid, HB, but to free $[H^+]$. The greater the concentration of hydrogen ions, the lower the value of pH, and vice versa. It should be remembered that if pH changes by 1 unit, the $[H^+]$ changes 10-fold, and if pH changes by 2 units, $[H^+]$ changes 100-fold. In other words, pH is an exponential function (Figure 1-3).

If the pH is known, the $[H^+]$ of any solution may be calculated by rearranging Equation 1-7 as follows:

$$\frac{1}{[H^+]} = 10^{\text{pH}} \qquad (1\text{-}8)$$

FIGURE 1-3
The relationship between pH and [H^+] from pH 6.7 to 8.0. Note that pH decreases exponentially as [H^+] increases. The dashed lines delineate the normal ranges for both parameters.

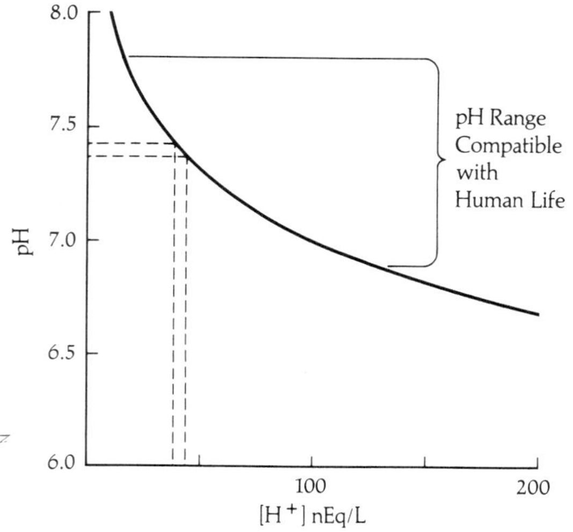

therefore,

$$[H^+] = 10^{-pH} = \frac{1}{10^{pH}} \tag{1-9}$$

For example, distilled water at room temperature (24°C) is said to be neutral, that is, neither acid nor alkaline. The pH of this water is 7.0. The [H^+] is calculated as follows:

$$[H^+] = \frac{1}{10^7} = \frac{1}{10000000} = 0.0000001\ Eq/L \tag{1-10}$$

The hydrogen ion concentration is one ten millionth of an equivalent per liter or 100 nEq/L.* The pH of arterial plasma in normal healthy adults is 7.40 (Table 1-2). The [H^+] of this plasma is 40 nEq/L. The pH values of extra cellular fluid (ECF) compatible with life, range from 6.9 to 7.8. The corresponding hydrogen ion concentration ranges from 126 to 16 nEq/L (Figure 1-1)†

*A nanoequivalent (nEq) is 10^{-9} equivalent. A microequivalent (μEq) is 10^{-6} equivalent. A millequivalent (mEq) is 10^{-3} equivalent. Nanoequivalents are used because they save writing a lot of zeros. For pure distilled water at 24°C, $[H^+] = 10^{-7}$ Eq/L $= 10^{-4}$ mEq/L $= 10^{-1}$ μEq/L $=$ 100 nEq/L.

† The definition of pH is actually the negative logarithm of the hydrogen ion activity. The activity is equal to concentration multiplied by the activity coefficient: $a = \gamma \cdot C$, where a is activity, C is concentration, and γ is the activity coefficient. In this text γ is assumed to be 1. Although this assumption is not strictly true for body fluids, it will not alter the important concepts presented in this text. A detailed discussion of activity coefficients may be found in textbooks of physical chemistry.

Table 1-2 Acid-Base Values in Adults and Children

(A) Adults

	pH	P_{CO_2} torr	$[HCO_3]$ mEq / L	Total CO_2 mmol / L
Arterial blood	7.37 – 7.43 (7.40)*	37 – 43 (40)	23 – 25 (24)	25 – 27 (26)
Capillary blood	7.39 – 7.44 (7.42)	36 – 42 (39)	22 – 26 (24)	22 – 27 (25)
Arterialized peripheral venous blood†	7.37 – 7.43 (7.40)	39 – 42 (41)	24 – 26 (25)	25 – 27 (26)
Peripheral venous blood	7.32 – 7.39 (7.37)	44 – 55 (49)	25 – 29 (27)	26 – 30 (28)

(B) Normal Acid-Base Values in Pregnancy and Childhood‡

		pH	H^+ nEq / L	P_{CO_2} torr	HCO_3 mEq / L
Pregnancy (arterial blood)	8 – 19 weeks	7.47 ∓ 0.02	34 ∓ 2	34 ∓ 2	23.8 ∓ 1.2
	20 – 29 weeks	7.48 ∓ 0.01	33 ∓ 1	32 ∓ 2	23.0 ∓ 1.7
	30 weeks	7.47 ∓ 0.03	34 ∓ 2	29 ∓ 3	20.4 ∓ 3.2
Childhood (capillary blood)	6 hr	7.38 ∓ 0.04	42 ∓ 4	37 ∓ 4	21.4 ∓ 1.9
	24 hr	7.41 ∓ 0.04	39 ∓ 4	37 ∓ 4	21.7 ∓ 1.9
	72 hr	7.42 ∓ 0.04	38 ∓ 4	36 ∓ 5	22.2 ∓ 3.1
	3 mos – 2 years	7.40 ∓ 0.03	40 ∓ 3	34 ∓ 4	20.1 ∓ 1.9
	1 – 3 years	7.38 ∓ 0.03	42 ∓ 3	34 ∓ 4	19.5 ∓ 1.4
	3 – 15 years	7.41 ∓ 0.03	39 ∓ 3	37 ∓ 3	22.7 ∓ 1.4

**Numbers in parentheses are mean values.*

†Arterialized peripheral venous blood is obtained by increasing blood flow in the peripheral veins of an extremity usually by heating the extremity.

‡All values are mean ∓ SD.

Source: *Acid / Base*, by J. J. Cohen and J. P. Kassirer. Copyright © 1982 by Little, Brown and Company. Reprinted by permission.

Blood Gases In the strictest sense, *blood gases* are the gases oxygen (O_2), CO_2, and nitrogen (N_2) found in blood. However, in clinical usage the term has evolved to refer to the P_{O_2}, P_{CO_2}, pH, and $[HCO_3^-]$. In other words, blood gases refer to the variables determined from blood–gas analysis. Blood gases are also referred to as blood–gas composition.

Buffer A *buffer* is a substance that lessens or absorbs *some* of the shock of a sudden change or impact. It does not prevent the change from occurring. In acid–base chemistry we say that a solution is buffered when it contains substances that lessen or reduce some of the change in pH that occurs when acids or alkalis are added to that solution. These substances are called chemical buffers. In other words, buffers must be able to absorb some of the excess H^+ or give up H^+ to a solution in which a paucity of hydrogen ions exists. If a buffer gives up H^+ to a solution, that is, dissociates H^+, it must be an acid. If a buffer combines with H^+ in a solution, it must be a base. Buffers are the conjugate acid–base pair (buffer pair) of weak acids in solution, that is, HB and B^- in Equation 1-3.

BODY FLUID COMPARTMENTS

The major constituent of the body is water. It is the solvent in which all solutes are either dissolved or suspended. At birth, water constitutes 75 to 80% of the body weight. However, by the time an infant is 1 year of age, total body water has decreased to about 65% of body weight. This percentage does not change until puberty. After puberty, body water constitutes about 60% of body weight in young adult males and 50% of body weight in young adult females. The difference is due to a greater percentage of adipose (fat) tissue in the adult female. Adipose tissue contains little water, hence the more fat tissue present in an individual, the lower the percentage of body weight that is water. By the time an individual reaches the seventh or eighth decade of life, the percentage of body weight that is water has decreased from the young adult values by about 10% in the male and about 6% in the female.

Total body water is divided into four principal compartments (Figure 1-4). The largest compartment is the intracellular fluid (ICF), which comprises 40% of body weight. Because of the larger muscle mass in males, the percentage of body weight that is ICF is somewhat greater than that of the female. The composition of ICF is shown in Table 1-3.

The second largest compartment is the ISF. This compartment constitutes about 15% of total body weight in both males and females. All cells of the body are bathed in an internal "sea" of ISF. Claude Bernard (1813–1878)

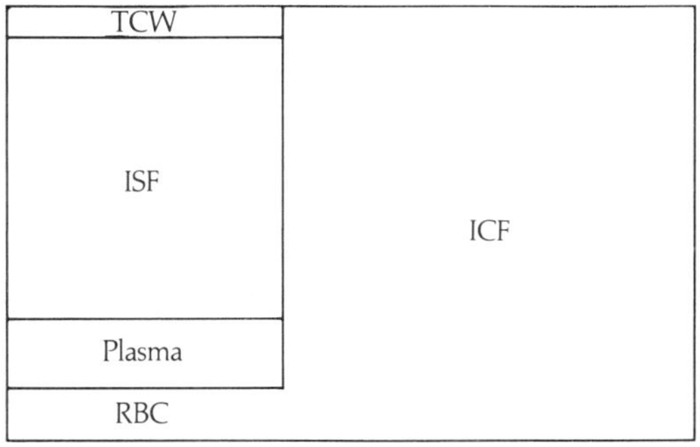

FIGURE 1-4
Body fluid compartments. Red blood cells (RBCs) are a part of the ICF, and this portion is shown as an extension of the ICF in this model. The plasma and RBC components constitute blood volume.

Table 1-3 Composition of Extracellular and Intracellular Fluid*

Cations and Anions, mEq / L	Plasma	ISF†	ICF
Na^+	140	145	(7 – 30)
K^+	4 (3.5 – 5.0)	4.1	(133 – 166)
Ca^{++}	5‡ (4.5 – 5.5)	3.4	(0 – 4)
Mg^{++}	1.6 (1.4 – 1.8)	1.3	(6 – 35)
pH	7.4 (7.37 – 7.43)	7.4	(6.9 – 7.23)
Anions			
Cl^-	102 (96 – 106)	118	(4 – 6)
HCO_3^-	25 (23 – 27)	28	(12 – 18)
Protein	15	0	(30 – 55)
$H_2PO_4^-$ – $HPO_4^=$	2.2	2.3	(4 – 40)
Other	6	5.5	(10 – 90)

**Ranges for plasma are shown in parentheses beneath mean values for each kind of ion. For ICF only ranges are given because each tissue type has its own characteristic composition and, thus, no single mean value is appropriate.*

†Plasma and ISF do not have exactly the same composition due to the presence of the protein anion in plasma, which creates a Donnan equilibrium across the capillary membrane. See Appendix 4.

‡Only 2.5 to 3 mEq / L are freely ionized in plasma. The remaining Ca^{++} is bound to protein. The concentration of calcium in milligrams per deciliter is twice the concentration in milliequivalents per liter.

called the ISF the internal milieu or internal environment of the body. The composition of ISF is markedly different from that of the ICF. For example, the principal cation of ISF is Na^+ whereas that of ICF is K^+. Both bicarbonate and chloride are found in higher concentrations in ISF than in ICF. Other differences are shown in Table 1-3.

The third compartment is the plasma. Plasma volume constitutes about 4% of body weight in the adult. In a sense, plasma is the "interstitial fluid surrounding blood cells." However, it differs in composition from true ISF as shown in Table 1-3. The main difference is the much higher protein concentration of plasma.

The smallest compartment, the transcellular water (TCW), constitutes only 1 to 2% of body weight in the normal individual. Some authors classify this compartment as part of the ISF but others consider it a separate compartment. In this text, TCW is considered a separate compartment. Transcellular water includes the fluid found in joints, the peritoneal and pleural cavities, the cerebrospinal fluid, fluid in the chambers of the eye, and secretions of the gastrointestinal system. The sum of the volumes of plasma, ISF, and TCW makes up the ECF, which is about 20% of the body weight.

The compartments present a dilemma for assessment of fluid–electrolyte and acid–base status. Because each has its own normal values, which of the four should be used for assessment? There are three criteria that must be met. First, a compartment that represents the body as a whole should be sampled. Second, the compartment must be accessible. Third, the sample must be obtained relatively easily with minimal risk and pain to the patient.

Initially, one might argue that ICF best represents the body as a whole. However, inspection of the data in Table 1-3 leads to the conclusion that there is considerable variation in composition between cell types. Also, sampling ICF that is not contaminated with blood and ISF is virtually impossible.

On the other hand, because ISF surrounds all living cells in the body it should be representative of the whole body. However, the ISF is not necessarily of uniform composition in all tissues. It is unlikely, for example, that ISF of skeletal muscle would have the same pH, P_{CO_2}, or $[HCO_3^-]$ as ISF from heart or brain tissue. Furthermore, it is nearly impossible to obtain pure samples of ISF from any source. Therefore, it must be concluded that neither ICF nor ISF are suitable fluids to sample for clinical assessment of acid–base status.

Plasma is a unique fluid in the body. It circulates through all tissues and the gases and electrolytes equilibrate with the ISF in each of those tissues. After equilibration with ISF in systemic capillaries, the plasma mixes in the right ventricle of the heart. After mixing, plasma has a composition that reflects an "average" composition of ECF in the body. It has also been shown that plasma obtained from a peripheral vein has essentially the same blood–gas and electrolyte composition as mixed venous blood flowing from the right ventricle. Therefore, plasma meets the first criterion of representing the body as a whole. The second and third criteria are also met because the plasma compartment is easily accessible and plasma samples can be obtained relatively easily with a minimum of risk and pain to the patient.

For acid–base assessment, the plasma compartment is sampled by drawing a small volume of blood from a vein or an artery and the sample is then exposed to O_2, CO_2, and pH electrodes. It should be noted that only pH of plasma is measured by this method because the pH electrode is exposed to the plasma and not to the ICF of the red cells. The P_{O_2} and P_{CO_2} are the same for both plasma and red blood cells (RBCs) because these gases equilibrate to the same partial pressure in plasma and red cell water. The $[HCO_3^-]$ is calculated from the P_{CO_2} and pH, hence the $[HCO_3^-]$ that is calculated applies to the plasma but not to whole blood.

For assessment of electrolyte concentrations, a sample of blood is drawn into a syringe and then transferred to a tube to clot. When the clot retracts, serum is separated from the red cell clot mass. The serum is then analyzed for electrolyte composition. Occasionally, a sample of plasma rather than serum is required for analysis. When plasma is to be analyzed instead of serum, the blood sample is placed in a tube containing an anticoagulant such as heparin. The results of analyzing serum or plasma should be the same because serum is essentially plasma without fibrinogen. Plasma is used when there is concern over the accuracy of the K^+ concentration measured in serum. This is discussed in more detail in Chapter 8.

BIOLOGICAL MEMBRANES

The fluid compartments are separated from one another by different membranes. RBC contents are separated from plasma by the erythrocyte membranes. Plasma is separated from ISF by the "capillary membrane" and ISF is separated from the intracellular compartment by cell membranes. Transcellular water is separated from other compartments by epithelial membranes.

Membrane Structure

Cell Membranes Cell membranes are made up of two layers of phospholipid plus protein. A phospholipid is a type of lipid (or fat) that is similar in structure to triglycerides (or neutral fats). A comparison of the structure of triglycerides and phospholipids is shown in Figure 1-5. Note that in triglycerides all three hydroxyl groups of the glycerol molecule combine with fatty acids. In phospholipids, the third hydroxyl group combines with phosphate, which is then combined with choline. The phosphate and choline make the phospholipid much more polar than triglycerides. The polar portions of the molecule can dissolve in water whereas the nonpolar part is lipid soluble.

The polar ends of the phospholipid make up much of the surface of the cell membrane. The nonpolar end of the molecule, which is made up of the long-chain fatty acid, is oriented perpendicular to the surface of the membrane (Figure 1-6). Thus, the center of the bimolecular cell membrane contains chains of fatty acids and gives the cell membrane its lipid matrix. The polar ends of the phospholipid make up the interior and exterior surface of the cell membrane. Cholesterol molecules (Figure 1-5) are also dissolved in the lipid matrix and oriented so that the polar hydroxyl group interacts with the polar end of the phospholipid molecule.

Proteins are interspersed in the almost fluidlike lipid layers. Some of the protein molecules are an intrinsic part of the membrane itself whereas others are attached only to the outer surface (Figure 1-6). It is thought that certain protein molecules are arranged so that they form aqueous channels (pores) through the lipid matrix. It is through these pores that small water-soluble solutes gain entry to or leave from the intracellular environment. Some water-soluble solutes must gain entry into the cell but are too large to pass

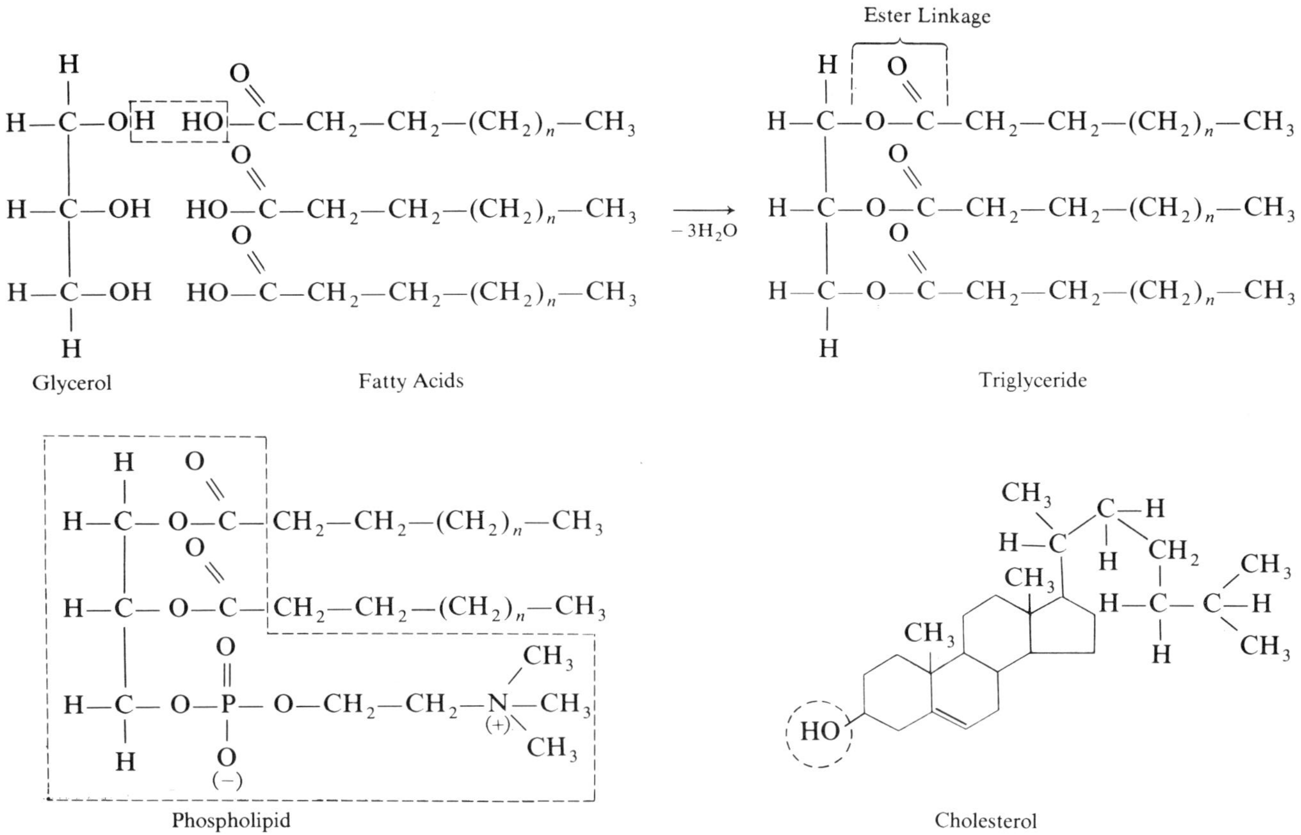

FIGURE 1-5 The structures of triglycerides, phospholipids, and cholesterol. Triglycerides are formed by combining fatty acids with glycerol (esterification). The value of *N* is typically 12, 14, or 16. In phospholipids one of the fatty acids is replaced with choline and phosphate. Note the charged portions of the choline and phosphate groups. These make the phospholipid molecule polar. Cholesterol has only one polar hydroxyl group. Those portions circled are the polar groups found on the surface of the cell membrane (Figure 1-6). The nonpolar parts make up the lipid matrix of the membrane.

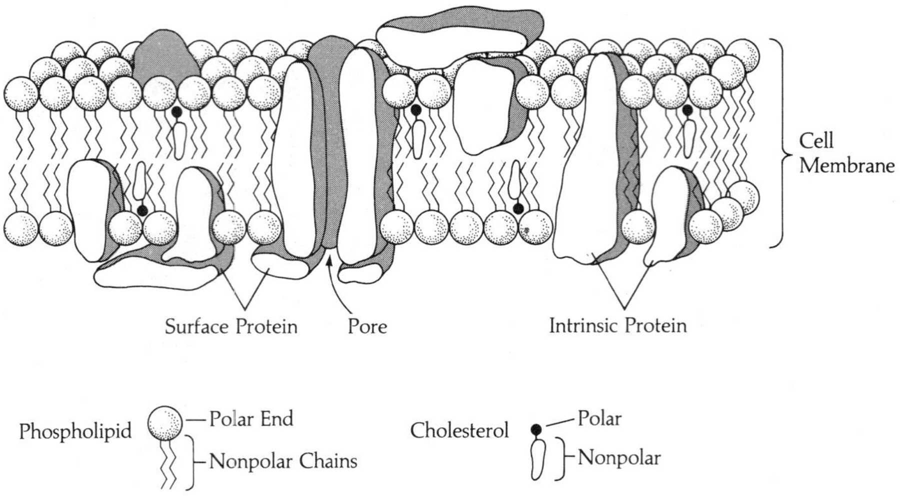

FIGURE 1-6 Model of the structure of a cell membrane. The long chains extending from the circular head of the phospholipid molecule are fatty acids. The choline portion of the molecule is a part of the polar head. Pores in the membrane probably are channels lined with protein. Cholesterol molecules are interspersed among the chains of fatty acids with the polar hydroxyl group oriented out toward the surface of the membrane.

through the pores. Some of the proteins probably function as carriers to transport these large molecules across the cell membrane. Other proteins serve as points of attachment for contractile elements, receptors, and enzymes on the cell surfaces.

Capillary Membrane The capillary membrane is a cylinder of squamous epithelial cells (called endothelium) that forms a barrier between the blood and the ISF (Figure 1-7). Pores are present in the membrane that permit passage of

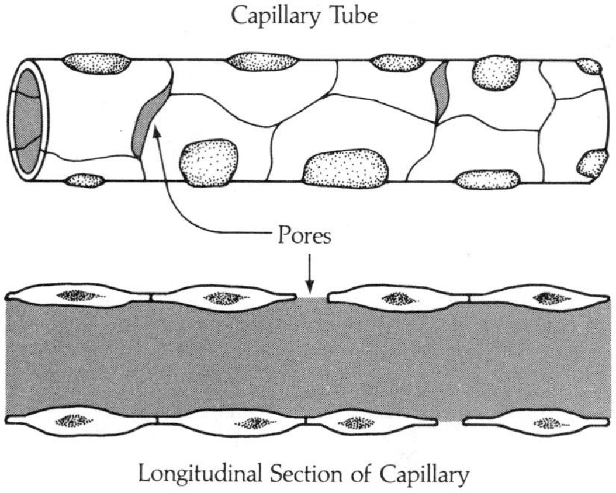

FIGURE 1-7
An example of endothelial membrane. The capillary tube has pores that may be slitlike, as shown in this diagram, or round and covered with a thin porous membrane. Endothelial membranes tend to be leaky and allow easy passage of water, crystalloid solutes, and gases.

crystalloid solute molecules and water. Small amounts of albumin and other proteins can cross through the pores; however, in health most of the protein is restricted to the plasma compartment. The capillary membrane is very important in fluid balance between plasma and ISF compartments. Details of capillary function are presented in Chapter 2.

Absorptive Epithelial Membranes Absorptive epithelial membranes separate TCW from ISF and plasma. These membranes include the mucosal epithelium of the stomach, intestine, and gallbladder, the pleural, peritoneal, and synovial membranes, as well as the tubules of the kidney. Renal, intestinal, and gallbladder epithelial membranes are specialized for absorption. Their general structure is shown schematically in Figure 1-8. The luminal cell surface of the epithelial membrane of the gastrointestinal (GI) tract and kidney is composed of microvilli which serve to increase surface area. This type of

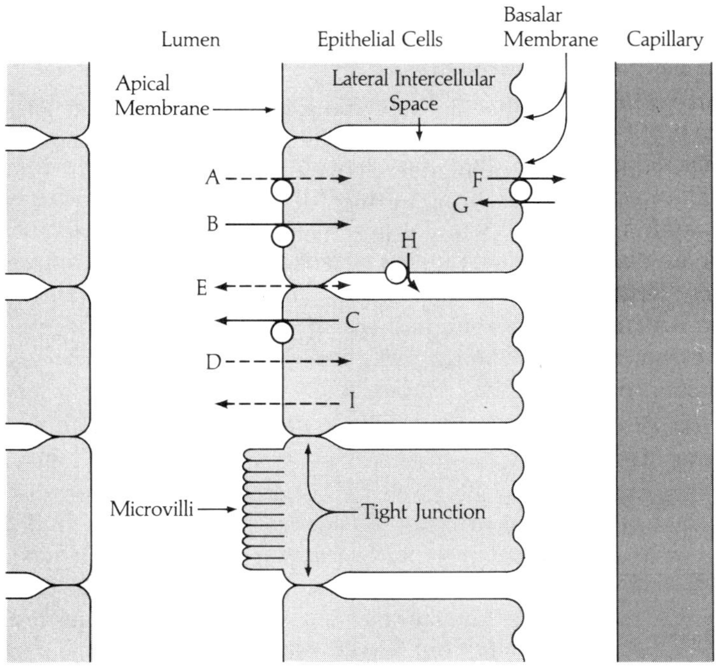

FIGURE 1-8 Absorptive epithelial membrane. Several different transport processes occur across these membranes. Special carriers may transport substances into the cell (A, B, and G), or out of the cell (F, H, and C). Substances may diffuse into (D) or out of the cell (I) or through tight junctions in either direction (E). In general, transport of substances into the lumen is called secretion, and transport from the lumen to the basalar side is called absorption. In the kidney, transport from the lumen to the ISF on the basalar side or to the intercellular space is called reabsorption. Many of these membranes have microvilli on the apical surface of the cell. To reduce confusion in the diagram, the microvilli are not shown on each cell.

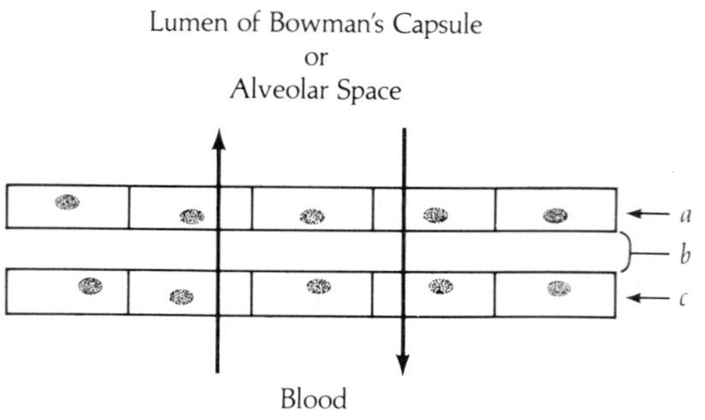

FIGURE 1-9 A schematic model of multicellular membranes. Two very important multicellular membranes are the glomerular membrane and the alveolar membrane. Each has a very specialized structure, however the commonalities are shown in this diagram. Cell layer *a* is a squamous epithelium that is in contact with alveolar air in the lungs or tubular fluid in Bowman's capsule in the kidneys. Layer *b* is a noncellular ground substance that contains some connective tissue elements and a "basement membrane." Cell layer *c* is endothelium of capillary. In the kidney, net transport is by means of bulk flow from blood into Bowman's capsule. In the lung, gases diffuse in both directions, O_2 into the blood and CO_2 from blood into the alveolar space.

membrane has carriers that actively transport solute. The carriers may be located in the membrane on either the apical (lumen) or basilar side of the cell. In the kidney tubule and gallbladder active transport also occurs in the lateral intercellular space (Figure 1-8). Further discussion of membrane transport function is included in the section titled "Transport Processes."

Multicellular Membranes Some membranes are more than one cell layer thick. Important examples are the alveolar membrane in the lungs and glomerular membrane in the kidneys (Figure 1-9).

Membrane Function

Membranes have several functions. First, they separate body fluid compartments from one another, in other words they serve as barriers between compartments. For example, proteins are kept inside of cells and ordinarily do not leak out across the cell membrane. The capillary membrane restricts the distribution of albumin and red blood cells to the plasma compartment. However, membranes do not completely isolate compartments from one another. Solutes such as glucose, oxygen, and amino acids cross the capillary membrane into the ISF and then cross cell membranes to meet cellular requirements for energy. Waste products produced by cell metabolism travel in the opposite direction for elimination from the body. Thus, membranes are selective in what they permit to cross from one compartment to the next, that is, they are selectively permeable to various solutes. Most membranes do not present a barrier to water transport between compartments, hence, water equilibrates between compartments.

A second major function of membranes is to provide structural sites for attachment of proteins. This function applies primarily to individual cell membranes. For example, proteins are attached to both the outer and inner surfaces of cell membranes. These proteins have several functions. They can serve as *receptors* for chemical messengers that help regulate cell metabolism and function. They provide *anchoring sites* for contractile proteins inside of cells; thus, when muscle proteins contract, the tension developed is applied directly to the cell membrane. In addition proteins located in membranes may act as *enzymes* for chemical reactions both inside and outside the cell. Finally, proteins may serve as *carriers* for transport of solutes into and out of cells.

TRANSPORT PROCESSES

Knowledge of transport processes is essential for understanding fluid–electrolyte and acid–base balance. Transport of fluids and electrolytes occurs between the body fluid compartments in health and during disturbances in homeostasis. Both acid–base imbalance and abnormal electrolyte composition in one compartment will affect fluid composition in other compartments by means of transport of fluid and electrolytes. The principles of transport discussed here apply to the fundamental processes by which fluids and solutes enter or leave a compartment. The concepts described are applied in later chapters to help explain both normal and disordered function.

Transport processes may be classified within two broad categories: bulk flow and nonbulk flow processes. These processes can be further broken down as follows:

—Nonbulk flow processes:
- Diffusion
- Carrier-mediated transport
  - Facilitated diffusion
  - Active transport

—Bulk flow processes:
- Filtration
- Osmosis

It is easier to understand bulk flow processes after learning concepts about nonbulk flow processes. Therefore, the discussion begins with nonbulk flow processes.

Nonbulk Flow Processes

Diffusion The simplest form of transport of atoms or molecules is diffusion. It is defined as the continuous random motion of molecules or atoms. Several factors affect the rate of diffusion, that is, how fast the molecules move from one location to another. These factors are the ambient (environmental) temperature, the molecular weight (or mass) of the substance, the distance over which diffusion occurs, the concentration of the substance, and the surface area available for diffusion. Each of these factors is discussed more fully in this section.

Energy for diffusion comes from heat (thermal energy) in the environment. The higher the ambient temperature, the faster the molecules will diffuse. Molecular size or mass also plays an important role in the rate of diffusion. Larger molecules diffuse more slowly at the same temperature than smaller molecules. For example, a molecule of glucose (molecular weight = 180) diffuses at approximately one third the rate of a molecule of water (molecular weight = 18).

Diffusion is reasonably efficient as long as distances are short. The greater the distance, the slower the rate of diffusion will be. Oxygen, for example, diffuses from the alveolar space into the blood in pulmonary capillaries. The thickness of the alveolar membrane between the alveolar space and plasma in the pulmonary capillary is about 200 nm. Because the distance is so short, diffusion is an adequate mechanism for transporting O_2 between alveoli and blood. On the other hand, if O_2 had to diffuse from the environment to all of the cells in the body, it would take years for the oxygen to reach the deepest tissues. Fortunately, once oxygen is in the blood, it is transported by blood flow to systemic capillaries where it diffuses into the ISF and ICF.

The rate of diffusion increases directly with the concentration of the substance; in other words, the greater the concentration of a substance, the faster the diffusion. Note in Figure 1-10 that initially all of the gas molecules are located on side 1 of the container. After a period of time, t, the number of molecules accumulating on side 2, will increase if the initial concentration of gas on side 1 is also increased.

Finally, the rate of diffusion increases in direct proportion to the surface area available for diffusion. Therefore, the greater the surface area, the greater will be the rate of diffusion for any given solute. The lungs have a very large surface area, 70 m^2. Disease processes, such as emphysema, that reduce this surface area will thus reduce the rate of diffusion of oxygen into blood.

The different factors affecting the rate of diffusion are related in the following equation:

$$\dot{q} = k_D A \frac{C}{L} \tag{1-11}$$

where $\dot{q}$ is the rate of diffusion (in moles or milligrams per unit of time), C is the concentration of the solute, A is the surface area available for diffusion, L is the distance over which diffusion occurs, and k_D the diffusion coefficient that contains factors such as temperature and molecular weight of the diffusing substance. The area term is not usually expressed as a separate variable but is included as part of the diffusion rate term. Then Equation 1-11 becomes

$$\frac{\dot{q}}{A} = J = k_D \frac{C}{L} \tag{1-12}$$

where J is the rate of diffusion per unit area and is called *flux*. Equation 1-12 expresses flux in one direction only. In Figure 1-10 the flux begins from side 1 to side 2 with no backflux from 2 to 1 because initially there is no gas on side 2. Therefore, the initial flux is proportional to concentration on side 1 (C_1). However, once diffusion begins, gas molecules begin to accumulate on side 2 and backdiffusion also commences. The same equation applies to diffusion in both directions except that for backflux the direction is opposite and, hence,

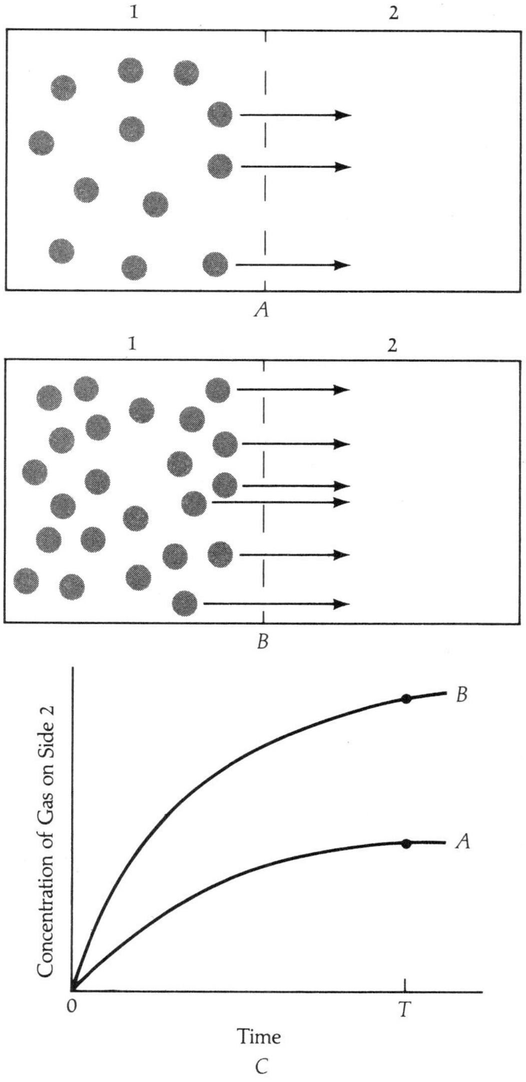

FIGURE 1-10
Simple diffusion. At time 0 in *A*, molecules diffuse across a membrane from compartment 1 to 2. In *B* at time 0, the concentration of molecules on side 1 is twice that shown in *A*. As diffusion progresses, the rate of accumulation of molecules on side 2 in *B* will be twice that of *A*. The relationship is shown graphically in *C*.

the sign is negative, as shown in Equation 1-13:

$$J = -k_D \frac{C_2}{L} \tag{1-13}$$

where C_2 is the concentration of gas on side 2. Net flux, or the rate of accumulation of gas on side 2, is the difference between flux from side 1 to side 2 and backflux from side 2 to side 1:

$$J_{net} = J_{1 \to 2} - J_{2 \to 1} \tag{1-14}$$

or

$$J_{net} = k_D \frac{C_1}{L} - k_D \frac{C_2}{L} \tag{1-15}$$

Because k_D and L are common to both terms in Equation 1-15,

$$J_{net} = k_D \frac{(C_1 - C_2)}{L} = k_D \frac{\Delta C}{L} \tag{1-16}$$

Equation 1-16 is called the Fick equation. The quantity $(C_1 - C_2)$ is the concentration difference (ΔC) between the two compartments. It should be noted that at equilibrium, the fluxes in both directions are equal, $C_1 = C_2$ and, therefore, under conditions of diffusion equilibrium net flux is zero.

When diffusion occurs across membranes, the Fick equation becomes modified to account for the presence of membrane factors that affect transport. First, solubility of the solute in lipid affects the rate of diffusion. Because cell membranes have a lipid matrix, the more lipid-soluble a solute, the quicker it will diffuse through the membrane. It is believed that a lipid-soluble solute first dissolves in the phospholipid matrix and then diffuses through the matrix to the opposite surface of the membrane. Other solutes that are not soluble in lipid, but are small enough, diffuse through the pores in the membranes. The relationships for solute diffusion across biological membrances are shown in Equation 1-17.

$$J = k_p \Delta C \tag{1-17}$$

The term k_p is the permeability coefficient, which contains all of the factors in the diffusion coefficient plus lipid-solubility factors and membrane thickness (L), which is the distance factor. ΔC is the concentration difference for the solute. The greater the value of the permeability coefficient, the greater will be the flux for a given ΔC.

Diffusion is the major transport process by which solutes enter and leave the plasma compartment in systemic and pulmonary capillaries. It is also the transport process by which O_2 enters plasma from the alveolar space and by which CO_2 leaves the plasma in the lungs. In addition, many drugs cross cell membranes or are absorbed into the blood by diffusion. Therefore, diffusion is a key transport process for maintaining homeostasis and delivering medications into the blood for circulation to tissues.

Carrier-Mediated Transport When substances are not lipid soluble and too large to diffuse through pores in the membrane, they are restricted to a given compartment (for example plasma protein) or they cross the membrane by means of carriers. The actual mechanism by which carriers bring about transport is not totally understood; however, the characteristics of carrier transport are well defined. Carriers are found in all cell membranes and are especially important in absorptive epithelial membranes.

There are three characteristics of carrier-mediated transport: saturation, specificity, and competition. Each characteristic is discussed in more detail in the following paragraphs and is illustrated in Figures 1-11, 1-12, and 1-13.

When all of the carriers for a specific solute become combined with that solute, the carriers are said to be *saturated*. Usually saturation occurs at higher than normal concentrations for a given solute. For example, glucose is transported across renal tubular cell membranes by a carrier (Figure 1-11). At normal glucose concentrations all of the glucose entering the tubule is reabsorbed. As glucose concentration increases in the tubular fluid, more of the

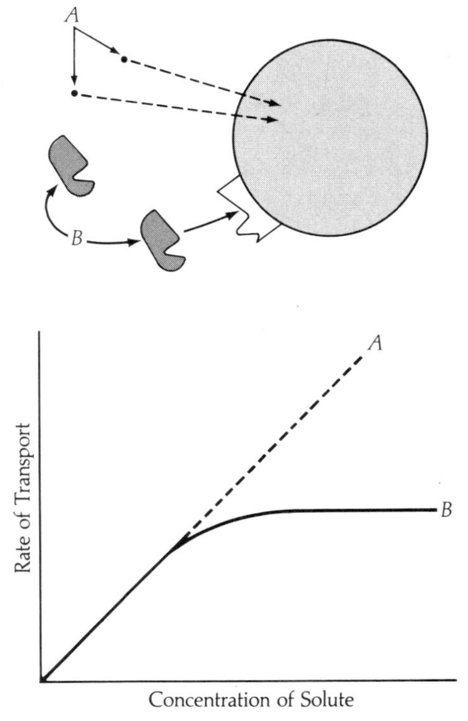

FIGURE 1-11
Saturation of carriers. Molecules labeled *A* cross the membrane by simple diffusion. Those labeled *B* enter by means of a carrier. When the concentration of both *A* and *B* increase, the rate of transport of each also increases. Molecule *A*, however, does not require a carrier to cross the membrane; hence, the rate of transport of *A* into the cell increases linearly as the concentration of *A* outside the cell increases. Molecule *B* requires a carrier to cross the membrane. When the carrier is saturated the rate of transport of *B* into the cell reaches a plateau, which is determined by the number of carriers transporting molecule *B*.

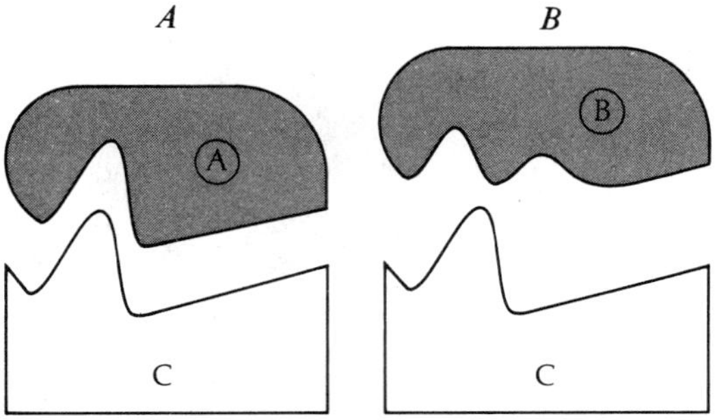

FIGURE 1-12
Specificity of carriers. Carrier *C* transports molecule *A* but cannot transport molecule *B* because *B* does not fit the carrier surface.

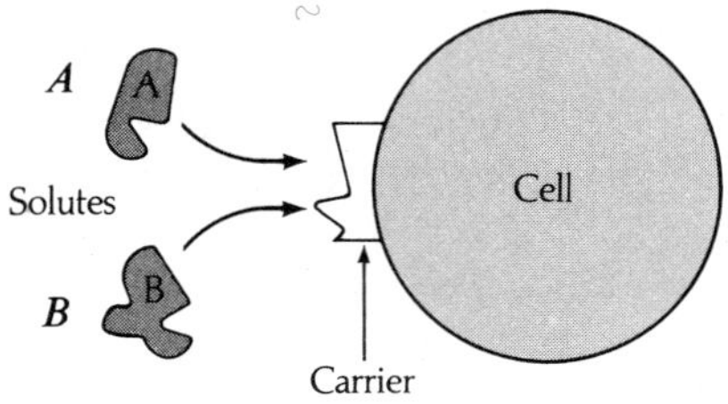

FIGURE 1-13
Competition for carriers. Both molecules A and B have similar structures and can combine with the same carrier on the cell membrane. The rate of transport of solute A when solute B is not present in solution will be greater than that rate of transport of A when B is present and competing for carrier sites.

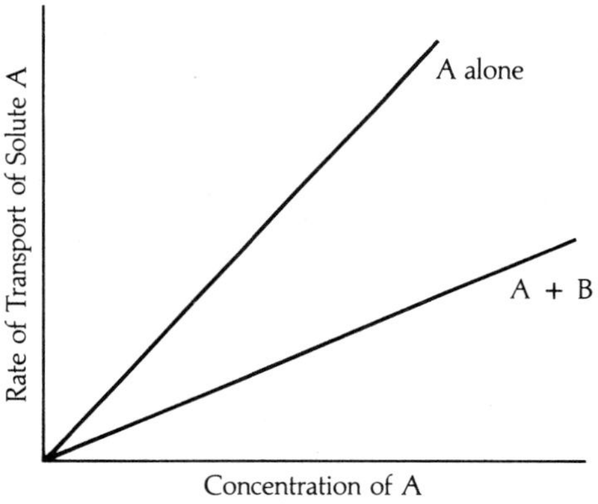

carriers become coupled with glucose. When more carriers combine with glucose, the rate of transport of glucose increases. However, when all of the carriers are combined with glucose (saturated), the rate of transport reaches a maximum. If glucose concentration of tubular fluid is increased further after the carriers are saturated, no further increment in glucose transport occurs. If glucose were transported across the renal tubular membrane by diffusion alone, no maximum rate of transport would be seen because there would be no carrier to saturate.

Carriers transport many different kinds of molecules. However, carriers that transport sugars such as glucose do not transport amino acids and vice versa.

This characteristic of carriers transporting only one kind of substance is called *specificity*. In some cases, the specificity is so complete that the carrier will combine with only one isomer of a given compound. The carrier for glucose, for example, transports only the D form of glucose and the L form will not combine with it (Figure 1-12).

Some molecules are similar enough in their structures that each can combine with the carrier. *Competition* for the same binding site on a carrier will occur when both molecules are being transported across a membrane (Figure 1-13). As a result, the rate of transport of either of the solutes will decrease when the other is present. Simple diffusion does not possess the characteristics of specificity or competition.

There are two kinds of carrier-mediated transport. The first is *facilitated diffusion* and the other is *active transport*. Both processes exhibit all three characteristics of carrier-mediated transport. Facilitated diffusion does not require energy from the cell for transport whereas active transport does require energy from the cell. With active transport solute is transported preferentially in one direction. Furthermore, active transport carriers or "pumps" can transport solutes against a concentration gradient—in other words, from compartments with low concentration of a solute to a compartment with high concentration of that solute. For example, cells actively transport K^+ from ISF where the concentration is only 4 mEq/L into the intracellular compartment where K^+ concentration is about 150 mEq/L. With facilitated diffusion the net transport occurs from the compartment with the higher concentration, but *net* transport does *not* occur against a concentration gradient. Glucose is transported across red cell membranes by facilitated diffusion. Both kinds of carrier-mediated transport processes are present in cell membranes and both play roles in normal regulation of acid–base and fluid–electrolyte balance. Specific examples are given in Chapter 4.

Solute Transport Across Absorptive Epithelial Membranes Solute transport across epithelial membranes (*transcellular transport*) involves both diffusion and carrier-mediated transport processes. The different kinds of transport processes can occur at either apical or basilar surfaces of the cell. The key point is that two separate membranes must be crossed or solutes must diffuse between the cells. Transcellular transport occurs across the intestinal epithelium in absorption of nutrients and across the renal tubular epithelium in the formation of urine. Examples of transcellular transport are discussed in Chapter 4 and a model is shown in Figure 1-8.

Bulk Flow Processes

Bulk flow is defined as the mass movement of either gases or fluid as a volume per unit of time in response to a pressure gradient. The pressure gradient may be a hydrostatic pressure gradient such as that produced by the contraction of the heart or a pressure gradient due to unequal concentrations of solutes across a membrane. Three kinds of bulk flow processes are discussed in this section: flow in a vessel or pipe, filtration, and osmosis.

Flow is the volume of gas or liquid moving in a tube or across a membrane per given unit of time. In a pipe or blood vessel, flow is produced by a hydrostatic pressure gradient. When pressure at one end of the vessel is greater than that found at the other end, flow is produced. The greater the *difference* in pressure, the greater will be the flow produced. When a liquid or gas flows through a pipe, energy is lost because of friction where the fluid contacts the vessel walls and also because of the collisions of molecules within the liquid (viscosity). Both of these factors reduce the energy contained in the fluid and impede the flow. The impediment to flow is called *resistance*, R. The relationship between flow, $\dot{Q}$, pressure difference, ΔP, and resistance is shown in Equation 1-18.

$$\dot{Q} = \frac{\Delta P}{R} \qquad (1\text{-}18)$$

It is clear from this equation that the greater the resistance, the smaller the corresponding flow will be. The components of resistance are the length of the vessel, the radius of the tube, and the viscosity of the fluid. The interrelationships of all of these variables are shown in Equation 1-19.

$$R = \frac{8\eta \cdot L}{\pi r^4} \qquad (1\text{-}19)$$

The Greek letter η represents viscosity or internal resistance of the fluid, L is the length of the tube, and r is the radius of the tube. Resistance increases directly with length of the tube and viscosity of the liquid. However, the resistance decreases in proportion to the fourth power of the radius of the tube. In others words, if the radius doubles, resistance decreases by 16-fold ($2^4 = 16$).

Flow of fluid across a membrane can be caused by a hydrostatic pressure difference or a solute concentration difference. Hydrostatic pressure differences force fluid through pores in the membrane, a process called *filtration*. The hydrostatic pressure in the systemic capillaries, for example, causes fluid to filter out through pores into the ISF. In the glomerular capillaries approximately 170 L of fluid are filtered per day. The equation for fluid flow in a pipe (Equation 1-18) may be modified for membranes, as shown in Equation 1-20 and also illustrated in Figure 1-14.

$$\dot{Q} = \frac{\Delta P}{R_m} \qquad (1\text{-}20)$$

where R_m is the resistance to flow through the membrane. The length term in membrane resistance is the length of the pore (or membrane thickness) and the radius is pore radius. For a given membrane, the flow (filtration) increases as the pressure difference increases.

FIGURE 1-14
Model for the process of filtration. Force applied to the piston on side *1* increases hydrostatic pressure on side *1* over that on side *2*. As a result, water is forced through the pores in the membrane from side *1* to side *2*.

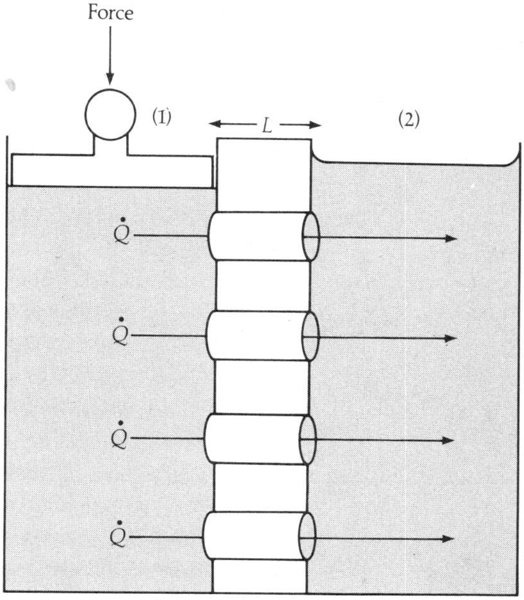

Osmosis is the flow of fluid across a semipermeable membrane against a solute concentration difference. A semipermeable membrane permits flow of water but not large solutes. A model of osmosis is shown in Figure 1-15. A nonpermeant solute is dissolved in solution on side 1 of the semipermeable membrane and a solution that does not contain the solute is contained on side 2. Initially water begins to diffuse out of the pore into the side containing the nonpenetrating solute (side 1). This diffusion of water occurs down its concentration gradient. In other words, the water in the pore is more concentrated than the water diluted with the nonpenetrating solute. The diffusion of water creates a decrease in pressure in the pore and water flows from side 2 into the pore because of the decreased pressure in the pore. The diffusion step occurs over an extremely short distance, probably no greater than the unstirred layer at the edge of the pore. Flow through the pore is the rate-limiting step and is a bulk flow process.

The greater the concentration of solute, the greater the rate of diffusion of water from the pore and the greater the corresponding osmotic flow. If pressure were applied to side 1 to prevent flow, the pressure that is required just to stop osmotic flow is called the osmotic pressure of the solution. The osmotic pressure may be calculated from Equation 1-21:

$$\Pi = CRT \tag{1-21}$$

where Π is the calculated osmotic pressure, C is the concentration of the nonpermeant solute in moles per kilogram of water (mol*al* concentration), R is the universal gas constant (0.082 L · atm/°K · mole), and T is the temperature in degrees (Kelvin scale). Since R and T are usually constant in body

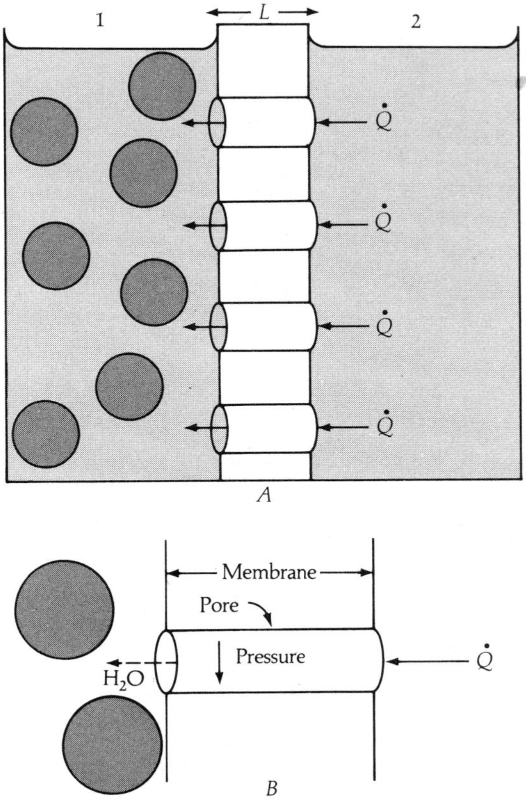

FIGURE 1-15
Model for osmosis. In *A* an impermeant solute is in solution on side *1* of the container. Only water is found on side *2* of the membrane. The solute dilutes the water on side *1*, so pure water diffuses from the pore (higher water concentration) into the solution on side 1 (lower water concentration). As a result, hydrostatic pressure in the pore decreases as shown in *B*. Water then enters the pore on side *2* because of the pressure gradient and flows to side *1*.

fluids, the concentration is the key variable. Frequently, more than one impermeable solute is present in solution. Under these conditions, the concentration term is given in units of osmoles per kilogram of water or osmolal concentration. An osmole provides 6.02×10^{23} solute ions or molecules, that is, 1 mole of solute particles regardless of charge. Interstitial fluid has a total solute concentration of approximately 0.3 osmol/kg H_2O, or 300 mosmol/kg water.

Osmotic flow may be calculated in a fashion similar to that for filtration:

$$\dot{Q} = \frac{\Delta\Pi}{R_m} \tag{1-22}$$

where $\Delta\Pi$ is the difference in osmotic pressure calculated for the solutions on the two sides of the membrane. R_m is the resistance to flow through the membrane. Frequently, osmotic flow and filtration occur across the same membrane. Under these circumstances the net flow is the algebraic sum of the osmotic flow plus filtration. This concept is discussed in detail in Chapter 2.

In the body, filtration occurs across capillary membranes, hence filtration is a mechanism by which fluid leaves the plasma compartment and enters the

ISF. Osmosis also occurs across capillaries from ISF to plasma. Osmosis occurs across all cell membranes when the solute concentrations on either side of the membrane change. However, generally there is no filtration across cell membranes.

SUMMARY

The body is composed of fluid compartments separated from each other by different kinds of membranes. Transport of solutes and fluid between compartments occurs as part of normal homeostasis. In general, diffusion occurs between all compartments, filtration occurs across capillary membranes, and osmosis can occur between any of the compartments. Carrier-mediated transport occurs across cell membranes and is an important component of transcellular transport. The section entitled *Fundamental Concepts and Definitions* should be reviewed before proceeding with subsequent chapters.

CHAPTER 2

Fluid and Osmolality Balance: Normal Homeostatic Processes

Both fluid and electrolyte balance are part of a larger set of homeostatic processes best described as material and energy homeostasis. In health there is a steady-state exchange of materials and energy with the environment called *external balance*. There is also a steady-state exchange of energy, fluid, and solutes between fluid compartments called *internal balance*. External and internal balance are connected through the circulatory, gastrointestinal, pulmonary, and renal (or urinary) systems (Figure 2-1). Because of the interconnections, if external exchange processes are not properly regulated, internal balance cannot be maintained within normal limits for very long. Thus, the organ systems participating in materials and energy balance must be regulated to maintain homeostasis. Normally, all the systems involved are very precisely regulated so that composition (see Table 1-3) and volume of body fluids are maintained within rather narrow limits.

EXTERNAL BALANCE

External or extrinsic balance includes the exchange of gases such as O_2 and CO_2, intake of food and other nutrients, water and electrolyte exchange, and heat exchange between the body and the environment. The several routes for these exchanges are shown in the diagram in Figure 2-1. Fluid and electrolytes enter the body at the oral end of the alimentary canal and are absorbed into the blood from the intestine. These substances leave the body through the

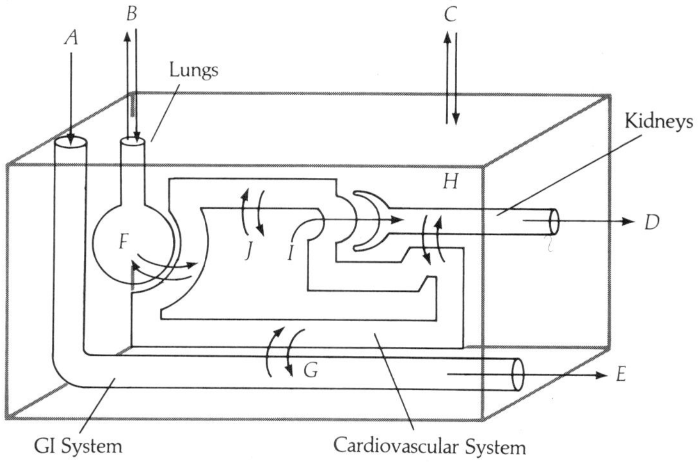

FIGURE 2-1 A model for exchange processes in the body. Water, food, and electrolytes enter the body at the oral end of the gastrointestinal (GI) system (*A*). Gases enter and leave the body via the lungs (*B*). Heat, water, and electrolytes are lost through the surface of the body (*C*), however, there is no significant gain of water and electrolytes through the integument. Excess water, electrolytes, and nitrogenous waste products of metabolism are excreted from the body through the kidneys (*D*). Lipid-soluble wastes are first metabolized by the liver and then excreted along with other organic wastes from the anal end of the GI system (*E*). A small amount of water is also lost through the excretion of feces. *F* represents exchange of gases between the blood and lungs, *G*, the exchange of nutrients, water, and electrolytes between blood and the GI tract, and *H* and *I* show the exchange of water-soluble solutes and water between the plasma and the kidneys. All of these routes are pathways in external balance. Internal balance is represented by *J*, which shows the exchange of solutes and water between ISF and plasma. Exchange of solutes and water between the ISF and ICF (not shown on this diagram) is also a part of internal balance.

kidneys, the anal end of the alimentary canal, and from the surface of the skin as sweat. Water is also lost from the body insensibly through lungs and skin.

To be in external balance with respect to a given substance, the total input or intake must equal the total output of that substance. For example, in sodium balance, the intake of sodium must equal output of sodium by way of all routes. Quantitatively,

$$\text{Input} = \text{Output} \qquad (2\text{-}1)$$

The normal intake route for sodium is oral (food and beverage) and normal output occurs through the urine, feces, and sweat. Therefore, Equation 2-1 may be modified as follows:

$$I = U_o + GI_o + S_o \qquad (2\text{-}2)$$

where I is oral intake, U_o is urine output, GI_o is gastrointestinal output, and S_o

Table 2-1 Average Values for Intake and Output of Water

Route	Average values, ml
Intake	
Beverages	1100
Food	950
Metabolism	350*
Output	
Insensible	800*
Sweat	50*
Fecal	150
Urine	1400

**These values are highly variable. They vary considerably with activity temperature and humidity. All of the values shown above are only averages in the most general sense; individuals can vary their intake of food and water over a wide range of values and still remain in water balance.*

is sweat output of sodium. This equation can be applied to potassium, chloride, and all other electrolytes as well as to sodium. When other routes of intake and output are involved, they must be added to the equation. For example, intravenous infusion of solutions containing sodium or other electrolytes constitutes an addition to input whereas loss of electrolytes from drainage through a nasogastric tube constitutes an addition to output.

For water balance, Equation 2-1 must be modified to include metabolic production of water. Production of water constitutes an input to the body stores of water (Equation 2-3).

$$\text{Intake} + \text{Production} = \text{Output} \qquad (2\text{-}3)$$

Water produced from metabolism comprises 12 to 15% of the total water input to the body. Output includes all of the water lost through the routes described previously, plus insensible loss. Average input and output values are shown in Table 2-1. It must be remembered that these values are averages only in the most general sense. Any of the subcomponents of intake and output can change markedly and balance may still be maintained. For example, oral intake of water may be doubled and balance will be maintained in the normal individual through increased urine flow. On the other hand, insensible loss may increase when an individual is exposed to a hot environment and be replaced by increased oral intake. Under these conditions, the kidneys also respond to increased water loss by concentrating the urine and thereby decreasing urinary water loss. In other words, there is considerable flexibility in the regulatory mechanisms that maintain water balance in health.

In disease flexibility for handling changes in water and electrolyte input decreases. For example, in chronic congestive heart failure the cardiovascular system may be easily overloaded and the kidneys may not receive sufficient perfusion to handle wide variations in solute and water input. Because of the decreased flexibility to respond adequately to solute and water loads, dietary restriction and medication may be needed to maintain a steady state. In congestive heart failure, a drug such as digitalis may be used to strengthen myocardial contractility and improve renal perfusion. In addition, a low-sodium diet (decreased input) and diuretic therapy (increased sodium and water output) may be used to help maintain external fluid balance.

INTERNAL BALANCE

Internal balance encompasses the steady-state exchange of energy, solutes, and water between body fluid compartments. Discussion of this topic is separated into two sections: (1) exchange between plasma and interstitial fluid (ISF), and (2) exchange between ISF and intracellular fluid (ICF).

Exchange Between Plasma and ISF

The major function of the circulatory system is to supply blood to the microcirculation so that essential nutrients, gases, products of cell metabolism, and wastes may be exchanged between plasma and ISF. It is through the exchange of these substances that the composition of ISF is regulated. In other words, exchange between plasma and ISF maintains the fluid environment in which cells live. All of the essential exchanges occur in the microcirculatory bed. This portion of the vascular system consists of several types of blood vessels including resistance vessels, capacitance vessels, exchange vessels, and shunt vessels (Figure 2-2). The exchange vessels are the capillaries and venules. All of the transport of water and solutes between ISF and plasma occurs across the walls of the exchange vessels.

Solute Exchange Between Plasma and ISF Net solute transfer across the exchange vessels occurs by diffusion. All capillaries are quite permeable to O_2, CO_2, and glucose. However, capillaries vary in their permeability to solutes such as HCO_3^-, H^+, and proteins. Some capillaries are very permeable. For example, the liver sinusoids possess large slitlike pores that permit diffusion of all solutes, including proteins such as albumin, across the exchange membrane. In the liver, protein can actually come into contact with the hepatocytes. This high permeability is necessary because the liver synthesizes most of the plasma proteins. These proteins can enter the plasma in the sinusoids. In addition, the liver metabolizes many protein-bound substances such as bilirubin and hormones. The high permeability permits the protein-metabolite complex (such as protein-bilirubin) to diffuse directly to the receptor site on the surface of the liver cells. The protein-metabolite complex then dissociates, leaving the protein in the plasma, and the metabolite combines with the receptor for transport into

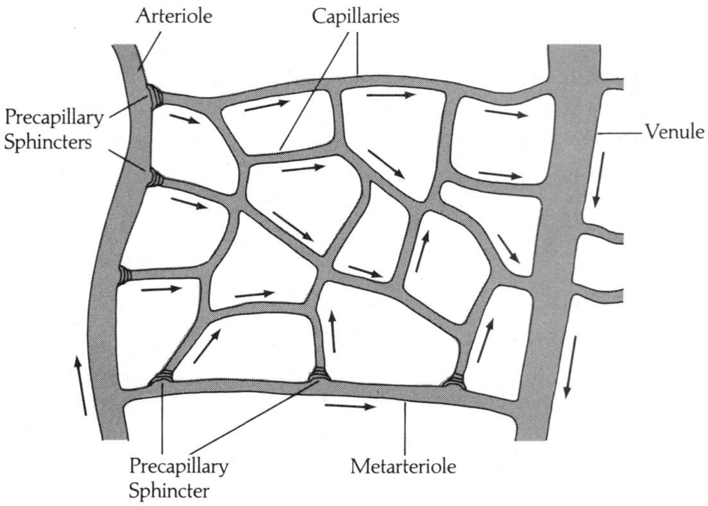

FIGURE 2-2 Diagram of the microcirculatory bed. The major resistance vessels are the arterioles. The metarteriole, present in some tissues, can supply blood to capillaries or serve as a direct channel to the venule. The venule is the main capacitance vessel in the network and serves with other venules and small veins as a reservoir of blood for the body. When venules and small veins contract, blood flow back to the heart increases. The capillaries are the major exchange vessels. These vary from 5 to 10 μm in diameter and from 0.5 to 1 mm in length. Direction of flow in the capillaries varies depending on whether precapillary sphincters are open or closed. The blood has been observed to flow in one direction for a short period and then reverse direction as a result of opening and closing precapillary sphincters. This phenomenon is called vasomotion.

the cell. Capillaries in brain tissue are at the other end of the permeability spectrum. These exchange vessels have very selective permeabilities to solutes. For example, HCO_3^- and H^+ equilibrate across those capillaries very slowly following changes in plasma concentration, whereas O_2 and CO_2 diffuse very quickly. Even urea does not equilibrate rapidly between the ISF of brain tissue and plasma. The structure of capillary pores in brain tissue probably contributes to the "blood-brain barrier." This barrier plays an important role in the respiratory response to acid-base disturbances. The transport of HCO_3^-, H^+, and CO_2 across the blood-brain barrier is discussed more thoroughly in Chapter 4.

The capillaries found in muscle, skin, the mesentery, and most organs have permeabilities between the two extremes described for liver and brain. Most crystalloid solutes and gases equilibrate rapidly across these exchange membranes. Normally, capillaries other than the liver sinusoids have low permeability to plasma proteins (colloids). There is a small leakage of protein out of capillaries, which is then returned to the plasma through lymph flow. It should be noted that permeability of the capillary to protein is increased during an inflammatory response. This change in permeability is part of the normal response to tissue injury.

Pulmonary capillaries have a higher permeability to protein than any other capillary bed except for liver sinusoids. Consequently, the volume of lymph removed from the lungs per day is also second only to that removed from the liver per day.

Water Exchange Between Plasma and ISF Water crosses the capillary and venular membrane by diffusion and bulk flow. The rate of diffusion of water across the exchange membrane is enormous and has been estimated to be 40 times greater than the volume of water brought to the capillary by the flow of blood. Therefore, water must diffuse rapidly back and forth through the pores in the membrane. The net result is neither gain nor loss of water in this process but rather an exchange.

However, there is a net loss of water from capillaries daily by bulk flow processes. Fluid flows between capillary plasma and ISF through a combination of filtration and osmosis. The mechanism of this exchange was described originally by Ernest Starling (1896) and is now known as the *Starling hypothesis*. A model of the ideal capillary, which is used to explain the Starling hypothesis, is shown in Figure 2-3. The capillary receives blood from an

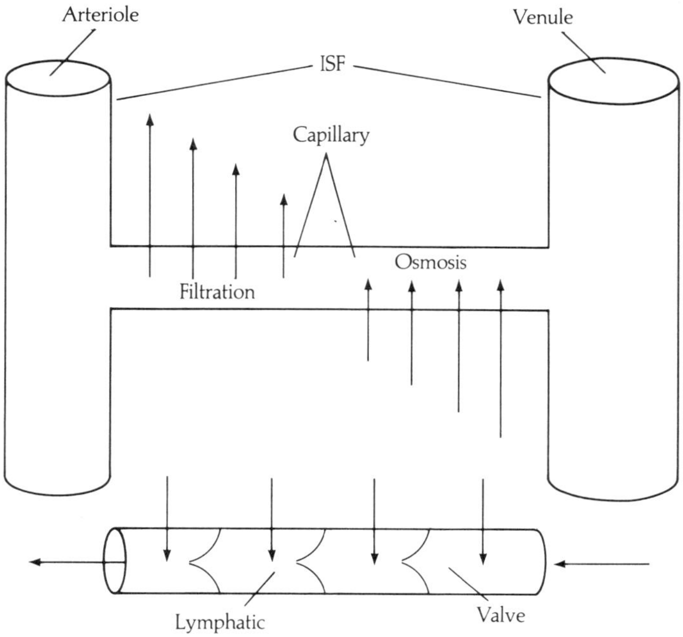

FIGURE 2-3 Model of the ideal capillary. Fluid is filtered out of the capillary into the ISF at the arteriolar end and reabsorbed by osmosis at the venular end. The length of the arrows indicate diminished filtration due to decreasing filtration pressure as blood flows along the length of the capillary. The reverse is true for osmosis. Excess fluid not reabsorbed is drained from the tissue via lymphatic vessels.

arteriole (a) and drains the blood into a venule (v). Hydrostatic (blood) pressure decreases from a high of about 25 mmHg at the arteriolar end to a low of approximately 10 mmHg at the venular end of the capillary. This pressure difference forces blood to flow from the arteriole along the length of the capillary to the venule. In addition, the hydrostatic pressure provides a driving force to filter fluid through the capillary wall. The pressure gradient for filtration is the difference between hydrostatic pressure in the capillary and that in the ISF (Equation 2-4).

$$\Delta P = P_C - P_{\text{ISF}} \tag{2-4}$$

P_C is capillary hydrostatic pressure and P_{ISF} is hydrostatic pressure in the ISF. The hydrostatic pressure of the ISF surrounding the capillary is approximately 7 mmHg less than atmospheric pressure.* Consequently, there is a hydrostatic pressure gradient favoring filtration along the entire length of the capillary.

There is also an osmotic pressure gradient between plasma and ISF. The concentrations of solutes in the ISF are similar to those of plasma except for protein (see Table 1-3). The protein in plasma makes the total solute concentration (osmolality) of plasma greater than that of ISF. The protein accounts for an osmotic pressure difference ($\Delta\Pi$) of about 25 mmHg (Equation 2-5). This osmotic pressure due to proteins is called colloidal osmotic pressure or oncotic pressure.

$$\Delta\Pi = \Pi_C - \Pi_{\text{ISF}} \tag{2-5}$$

Because of this gradient there tends to be osmosis from ISF into capillary plasma. Π_C is capillary oncotic pressure and Π_{ISF} is the oncotic pressure of ISF.

Because filtration and osmosis oppose each other, net flow across the capillary membrane is the difference between the two flows (Equation 2-6).

$$\dot{Q}_{\text{net}} = \dot{Q}_F - \dot{Q}_o \tag{2-6}$$

$\dot{Q}_F$ is flow due to filtration and $\dot{Q}_o$ is flow due to osmosis. In Chapter 1, flow due to filtration is given as

$$\dot{Q}_F = \frac{\Delta P}{R_m} \tag{2-7}$$

where ΔP is the hydrostatic pressure difference across the capillary membrane ($P_C - P_{\text{ISF}}$) and R_m is the resistance to flow through the membrane. Osmotic

*Not all investigators agree that the pressure in the ISF is subatmospheric. Whether the pressure is atmospheric (zero) or subatmospheric will not alter the concepts presented in this discussion.

flow was also described in Chapter 1 and is summarized in Equation 2-8:

$$\dot{Q}_o = \frac{\Delta\Pi}{R_m} \qquad (2\text{-}8)$$

where $\Delta\Pi$ is the osmotic pressure difference across the capillary membrane ($\Pi_C - \Pi_{ISF}$) and R_m is the resistance to flow through the membrane. Equations 2-7 and 2-8 may be substituted into Equation 2-6:

$$\dot{Q}_{net} = \frac{\Delta P}{R_m} - \frac{\Delta\Pi}{R_m} \qquad (2\text{-}9)$$

Since R_m is common to both terms on the right side of Equation 2-9, it may be factored out as shown in Equation 2-10.

$$\dot{Q}_{net} = \frac{1}{R_m}(\Delta P - \Delta\Pi) \qquad (2\text{-}10)$$

The term, $1/R_m$, is the conductance of the membrane and is usually referred to as the filtration coefficient, K_f. Thus, Equation 2-10 becomes

$$\dot{Q}_{net} = K_f(\Delta P - \Delta\Pi) \qquad (2\text{-}11)$$

Equations 2-4 and 2-5 may be substituted into Equation 2-11:

$$\dot{Q}_{net} = K_f[(P_C - P_{ISF}) - (\Pi_C - \Pi_{ISF})] \qquad (2\text{-}12)$$

Equations 2-11 and 2-12 are mathematical expressions of the Starling hypothesis. In a healthy person, K_f is usually constant, hence net flow across the capillary membrane is dependent on hydrostatic and osmotic pressure differences. The difference between the hydrostatic pressure difference (ΔP) and osmotic pressure difference ($\Delta\Pi$) is called net filtration pressure (NFP).

$$\text{NFP} = (\Delta P - \Delta\Pi) \qquad (2\text{-}13)$$

and, therefore

$$\text{NFP} = (P_C - P_{ISF}) - (\Pi_C - \Pi_{ISF}) \qquad (2\text{-}14)$$

When the hydrostatic pressure difference exceeds the osmotic pressure difference, net filtration pressure will be positive and net flow will be positive, that is, net flow will be out of the capillary due to filtration. In the model of the ideal capillary in Figure 2-3, net filtration pressure is positive throughout the first half of the capillary. When the osmotic pressure difference exceeds the hydrostatic pressure difference, net filtration pressure is negative and flow across the capillary membrane is negative, that is, from ISF into capillary plasma. In the model (Figure 2-3), net filtration pressure is negative in the distal half of the capillary.

Fluid exchange between ISF and capillary plasma probably does not follow the model in Figure 2-3 exactly. When arterioles and precapillary sphincters in the microcirculatory bed dilate, the pressure in the capillary increases so that filtration occurs along the entire length of the capillary (Figure 2-4). Dilation of the resistance vessels is then followed by a period of constriction of the resistance vessels. During the period of vasoconstriction, the hydrostatic pressure in the capillary is reduced, making net filtration pressure negative. As a result, osmosis occurs along the entire length of the capillary. The net result is that most of the fluid that is filtered is reabsorbed by osmosis. The excess fluid that is filtered is returned to the plasma by means of lymphatic drainage of the ISF.

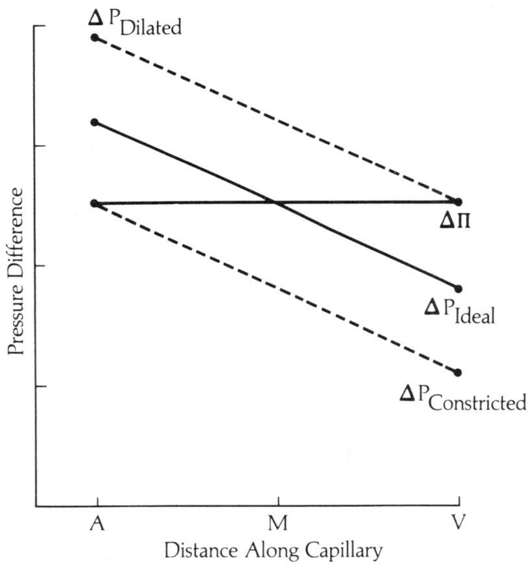

FIGURE 2-4
Differences in pressures between capillary blood and ISF along the length of the capillary. The hydrostatic pressure difference, ΔP, (Equation 2-4) diminishes as blood flows from the arteriole to venule. The oncotic pressure difference, $\Delta \Pi$, (Equation 2-5) remains relatively constant throughout capillary length. ΔP_{Ideal} is the hydrostatic pressure difference seen in the ideal capillary as depicted in Figure 2-3. $\Delta P_{Dilated}$ shows the hydrostatic pressure difference when arterioles and precapillary sphincters dilate and $\Delta P_{Constricted}$ shows the hydrostatic pressure difference when arterioles and precapillary sphincters are constricted. *A* indicates the arteriolar end, *M* the middle, and *V* the venular end of the capillary.

Lymph Vessels In the model in Figure 2-3 filtration is nearly balanced by the process of osmosis. Therefore, only a small fraction of the total flow leaving the capillary by filtration remains in the ISF. This excess fluid, along with the small fraction of protein that diffuses out of the capillary, is removed by lymphatic vessels. Excess fluid in the ISF enters the lymph capillaries by tissue pressure. Contraction of skeletal muscles facilitates entry of ISF into the lymphatic capillary as well as flow in the larger lymph vessels. Lymphatics possess many one-way valves similar to those found in veins, hence flow is one way through the system. The fluid drains from lymph capillaries to larger lymph vessels, which in turn ultimately connect to the right and left subclavian veins near the junction of the subclavian veins with the internal jugular veins. While in the lymphatic system, the fluid passes through lymph nodes that filter out foreign particles such as bacteria and foreign proteins that have escaped phagocytosis by macrophages in tissues. It should be noted that the body is richly endowed with lymph vessels.

The volume of lymph removed from tissues varies from 2 to 4 L per day. In addition one fourth to one half of the total protein content of plasma diffuses out of capillaries and is returned each day to the vascular space by lymphatics. Lymphatic vessels are the only route by which protein may be returned to the circulation. If protein did not return by way of lymphatics, then it would remain in the ISF and raise the oncotic pressure of ISF. When oncotic pressure of ISF is increased, net filtration pressure (Equation 2-14) will increase, which in turn can lead to edema. The mechanisms that lead to edema are discussed in Chapter 7.

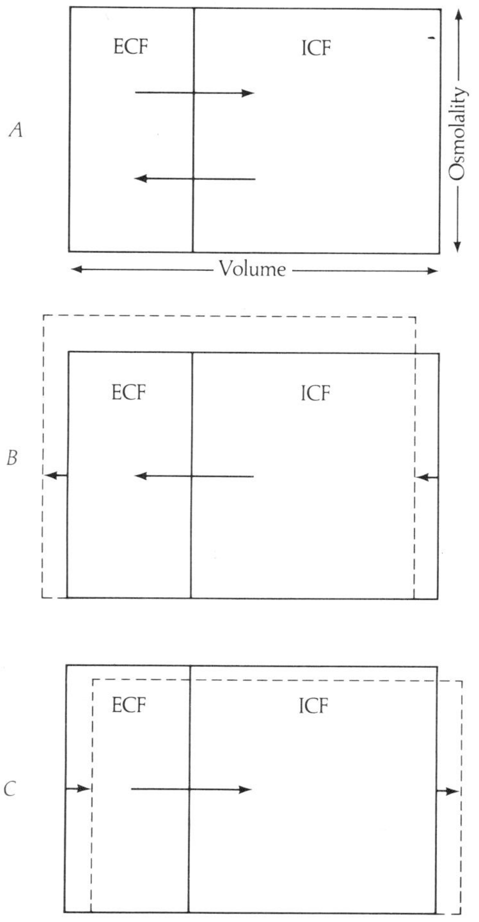

FIGURE 2-5
Fluid exchange between the intracellular and extracellular compartments. The volume of each compartment is indicated by its length and the osmolality by its height. In *A* osmotic equilibrium is present between the two compartments. In *B* osmolality of ISF, and therefore ECF, has increased causing osmosis of water from the ICF, which in turn expands extracellular volume. In *C* the osmolality of ISF has decreased and there is a concomitant osmosis of water from the ISF into the ICF until osmotic equilibrium is restored. This restoration of equilibrium results in a decreased volume of ECF. Note that the figure shows exchange between ICF and ECF, not ISF alone. It must be remembered that for fluid to shift into or out of the ICF, osmolality of the ISF must change. However, when osmolality of ISF changes, there will be a corresponding change in the osmolality of plasma. When volume shifts from the ICF, both the ISF and plasma volumes will increase. Hence, in this figure, gain and loss of fluid are shown as changes in volume of the ECF and not ISF alone.

Exchange Between ISF and ICF

The interstitial fluid and intracellular fluid are in osmotic equilibrium across cell membranes. Therefore, if the osmolality of one compartment changes, water flows by osmosis across the cell membrane until osmotic equilibrium is reestablished. For example, when the osmolality of the ISF increases compared to that of the ICF, water flows by osmosis from ICF to ISF until osmotic equilibrium is restored. The reverse is also true—if ISF osmolality becomes less than that of ICF, water flows by osmosis from ISF to ICF (Figure 2-5). Exchange of water between ISF and ICF occurs by osmosis, not filtration. Changes in hydrostatic pressure in the ISF are transmitted directly across cell membranes to the ICF so that pressure changes in the ICF parallel those in the ISF. Hence, filtration across cell membranes will not normally occur.

It should be noted that when fluid shifts from ISF to ICF in a person in external balance, the volume of the plasma compartment will decrease. Furthermore, if water shifts from ICF to ISF, some of that water will enter the plasma compartment and cause an expansion of blood volume. Thus, changes in ISF volume will alter the volume of the plasma compartment. Disorders in the balance between ISF and ICF are discussed in Chapter 7.

REGULATION OF THE OSMOLALITY OF BODY FLUIDS

Regulation of the osmolality of both intracellular and extracellular fluid is accomplished by regulating the intake (ingestion) and output (urinary excretion) of water. Central to the mechanism regulating water balance is a feedback control system that includes the kidneys, the hypothalamus, the posterior pituitary gland (neurohypophysis), and antidiuretic hormone (ADH). A diagram of the control system is shown in Figure 2-6.

Dehydration that causes increased osmolality and any factor that decreases plasma volume will stimulate nerve cells whose cell bodies are located in the supraoptic and paraventricular nuclei of the hypothalamus. The axons of these nerve cells terminate in the posterior pituitary gland. When stimulated, these neurons secrete the octapeptide, ADH, which in turn diffuses into the blood perfusing this gland. The ADH is then circulated to the kidneys, where it increases the permeability of the distal tubule and collecting duct to water. As a result, the kidneys reabsorb more water, thereby decreasing plasma osmolality and also reducing urinary water loss.

The hypothalamus also helps regulate ingestion of water through the thirst mechanism. Both increased osmolality and volume depletion stimulate thirst in normal people. Dryness of the mucous membranes in the mouth and pharynx also stimulate thirst and increase water ingestion. In fact, in many animals and to a limited extent in humans, the volume of water ingested to slake thirst will just replace water losses. This is rather remarkable considering that the

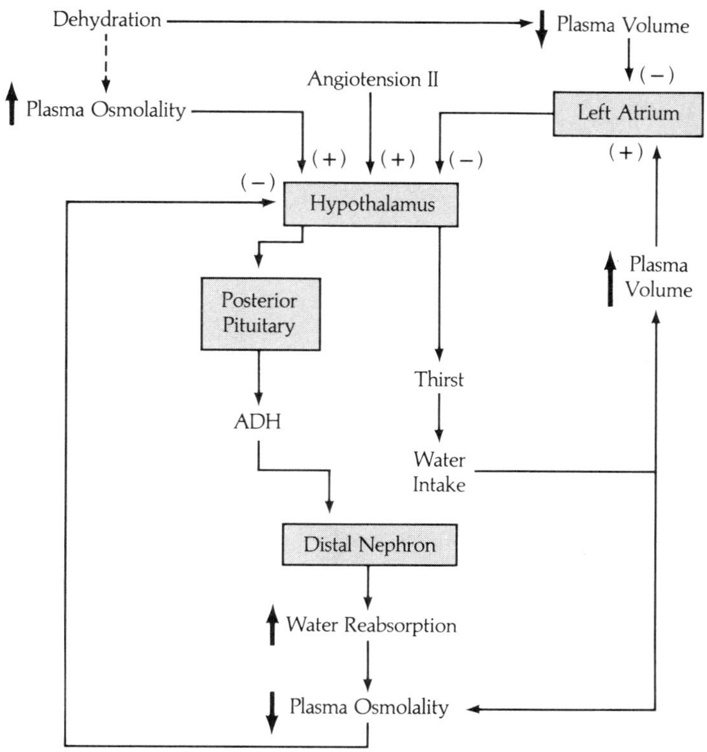

FIGURE 2-6 A model for feedback regulation of osmolality of the ECF. Angiotensin II and increased plasma osmolality stimulate (+) the hypothalamus to cause ADH secretion from the posterior pituitary gland. Distension of the left atrium and decreased plasma osmolality inhibit the hypothalamus, thereby reducing ADH secretion. Dehydration can cause increased plasma osmolality and decreased plasma volume. The latter results in decreased stretch of the left atrium and, thereby, diminishes inhibition of hypothalamic nuclei.

ingested water has not had time to enter the circulation and influence either plasma osmolality or volume. The mechanisms by which the volume being ingested is monitored are unknown. In general, increased water reabsorption by the kidney and water ingestion help reduce plasma osmolality. Water intake also contributes to the restoration of volume in fluid compartments.

There are two additional components to the feedback control of water balance. The first is a group of peptides called angiotensins. Angiotensins stimulate ADH secretion by directly stimulating the hypothalamus. Therefore, those factors that influence angiotensin production also influence secretion of ADH. Angiotensin is discussed further in the next section under *Regulation of Extracellular Volume*. The second component is the baroreceptor reflex arising from the left atrium of the heart. These baroreceptors are stimulated when increased pressure stretches the walls of the left atrium. When plasma volume or left atrial pressure increases, *inhibitory* nerve impulses are sent over afferent pathways to the paraventricular and supraoptic nuclei of the hypothalamus.

Therefore, increased left atrial pressure can reflexively decrease ADH secretion. Decreased pressure in the left atrium reduces the inhibition of ADH release.

REGULATION OF EXTRACELLULAR VOLUME

The regulation of the volume of extracellular fluid is enmeshed in the processes regulating sodium balance. The fundamental idea is that if sodium is retained, water will also be retained, and conversely, if sodium is excreted, water will also be excreted. Extracellular volume is remarkably stable in health; therefore, sodium balance must be regulated within narrow limits. Regulation of sodium balance is closely intertwined with regulation of other electrolytes and acid-base balance. Consequently, changes in sodium balance frequently upset fluid and electrolyte balance as well as acid-base status of the body.

Maintenance of extracellular volume is brought about by retaining or excreting sodium.* Intake must equal output of sodium to maintain a steady state. Sodium is found in most foods and is readily available as a dietary supplement. The average American ingests 20 to 30 times more sodium chloride (NaCl) than is needed to maintain balance. A normal person can easily maintain sodium content and volume of the extracellular fluid with as little as 0.5 g NaCl per day plus an adequate water supply. Generally, intake of sodium and water is more than sufficient for minimum body needs.

Regulation of sodium and water output or excretion involves a complex set of intertwined processes in the kidneys, adrenal cortex, and blood. The kidneys are the route of output for excess sodium and water, but these organs must be regulated to prevent either excessive or insufficient excretion. Three factors are known to be of major importance in regulating sodium and water excretion. These are (1) the glomerular filtration rate (GFR), (2) the hormone aldosterone, and (3) physical factors within the kidney itself.

Changes in GFR will alter sodium excretion. In general, increased GFR will lead to an increase in sodium excretion, and decrements in GFR will reduce sodium excretion. The physical factors in the kidney involve changes in protein concentration in plasma of the peritubular capillaries, which are important in balancing reabsorption (especially in the proximal tubule) with changes in GFR. This process, which is part of glomerulotubular balance, is described in texts of renal function but is tangential to the discussion at hand. Aldosterone, on the other hand, plays a major role in both fluid-electrolyte and acid-base regulation. Therefore, a short discussion of the effects and the regulation of aldosterone secretion is included here.

Aldosterone is a steroid hormone that is produced by the cells of the zona glomerulosa of the adrenal cortex. This hormone stimulates sodium reabsorp-

*To be absolutely correct, the sentence should read "... sodium and its accompanying anions." Anions are implicit whenever sodium is being discussed but will not be mentioned unless reference to a specific anion is needed.

tion and potassium secretion by the cells of the distal tubules and collecting ducts in the kidney. The amount of sodium reabsorption dependent on aldosterone activity is only 2% of the total amount of filtered sodium. Two percent of the filtered sodium is about 30 g per day, which is twice the high normal intake of sodium per day even in American diets. Without aldosterone this sodium would be lost in the urine. The rate of sodium reabsorption increases in the distal nephron as aldosterone concentration increases. At very high concentrations of aldosterone in plasma, the urine may be virtually free of sodium. When aldosterone concentration rises, sodium concentration in urine decreases while potassium concentration increases. The reverse occurs when aldosterone concentration decreases. It is of interest that aldosterone also stimulates sodium uptake from the intestine as well as from fluid in the lumen of ducts of sweat glands and salivary glands.

Control of aldosterone secretion is affected by at least four separate factors: (1) sodium concentration in plasma, (2) potassium concentration in plasma, (3) adrenocorticotropic hormone (ACTH), and (4) renin secretion from the kidneys (Figure 2-7). The first three are of less importance than renin secretion.

There are sodium receptors within the cells of the zona glomerulosa of the adrenal cortex. When sodium concentration in plasma increases, intracellular concentration of sodium is also thought to increase. The increased sodium concentration causes a reduced rate of aldosterone secretion and vice versa.

Increased potassium concentration increases aldosterone secretion rate. If sodium concentration of plasma falls at the same time potassium concentration rises, the effect on aldosterone secretion is additive. If the plasma concentration of both sodium and potassium increase, the net effect on aldosterone secretion is the algebraic sum of the two.

ACTH is secreted by the anterior pituitary gland in response to stressful stimuli. The main effect of ACTH is to stimulate synthesis and release of glucocorticoids such as cortisol from other regions of the adrenal cortex. When ACTH is secreted in high concentrations, for example, following physical trauma, it will cause secretion of a small amount of aldosterone. However, unlike the role of ACTH in regulating cortisol secretion, the role of ACTH in regulating aldosterone secretion is relatively minor. The structure and sequence of synthesis of cortisol and aldosterone are shown in Figure 2-8.

The major regulator of aldosterone secretion is an enzyme, renin, which is secreted into the blood by the granular cells of the juxtaglomerular apparatus of the kidney (Figure 2-7). The juxtaglomerular apparatus is the site where the junction between the ascending limb of the loop of Henle and the distal tubule makes contact with the afferent arteriole. The junction is called the macula densa. There are three types of cells in this apparatus: granular cells, which are closely associated with the wall of the afferent arteriole, mesangial cells whose functions are unknown, and macula densa cells. A diagram of the juxtaglomerular apparatus is shown in Figure 2-9. Renin catalyzes the conversion of angiotensinogen, a plasma protein synthesized by liver cells, to angiotensin I, a decapeptide. Angiotensin I is then converted to angiotensin II, an octapeptide, by splitting two terminal amino acids from the decapeptide. The enzyme

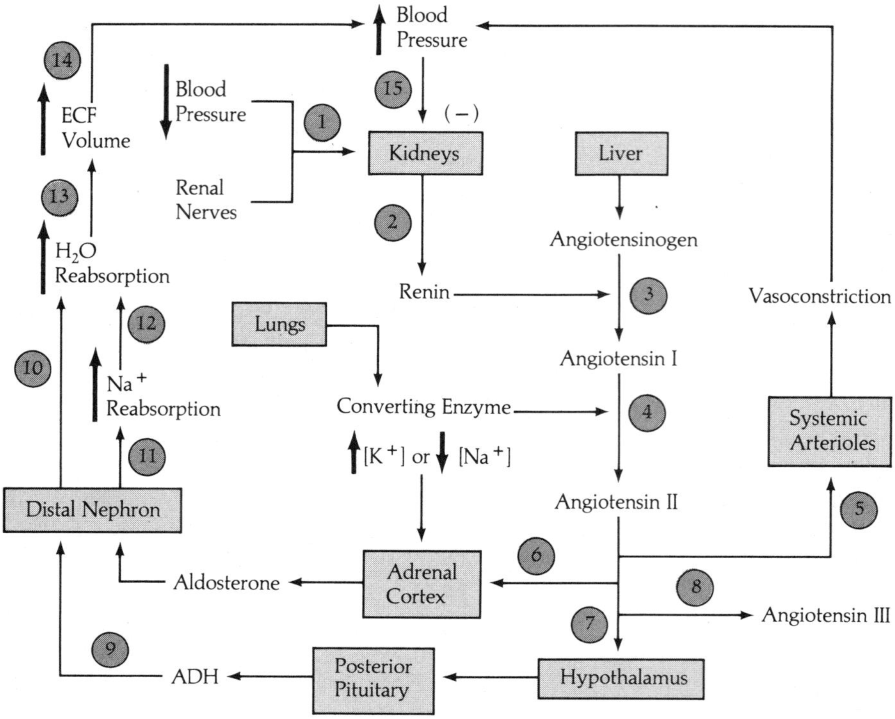

FIGURE 2-7 A model for the feedback regulation of extracellular volume and blood pressure. When blood pressure decreases in the kidneys or renal nerves are stimulated (1), renin is secreted into the blood (2). Renin hydrolyzes a plasma protein called Angiotensinogen into a decapeptide, angiotensin I. Converting enzyme in the lungs splits two terminal amino acid molecules from angiotensin I forming angiotensin II, an octapeptide. Angiotensin II causes vasoconstriction of systemic arterioles (5), stimulates the adrenal cortex to release aldosterone (6), and stimulates the hypothalamus (7) to cause ADH release from the posterior pituitary gland (9). Aldosterone causes an increased rate of sodium reabsorption by the distal nephron (11) and together with ADH (10 and 12) increases water reabsorption, which expands the ECF volume (13). Both vasoconstriction (5) and increased ECF volume (14) serve as negative feedback to the kidneys by increasing blood pressure (15).

responsible for this step is called converting enzyme and is found in the lungs and certain other tissues. Angiotensin II can be further split to a heptapeptide, angiotensin III (Figure 2-7). Angiotensin I appears to be biologically inactive, but II and III are active. Angiotensin II and possibly III stimulate the synthesis and release of aldosterone from the zona glomerulosa of the adrenal cortex. Angiotensin II also serves as a negative feedback to decrease renin secretion, and it stimulates secretion of ADH as described previously. Other details of the feedback control system for aldosterone secretion are shown in Figure 2-7.

Renin release is stimulated by several processes. Reduced blood pressure in the kidney will bring about renin release whereas increased pressure decreases

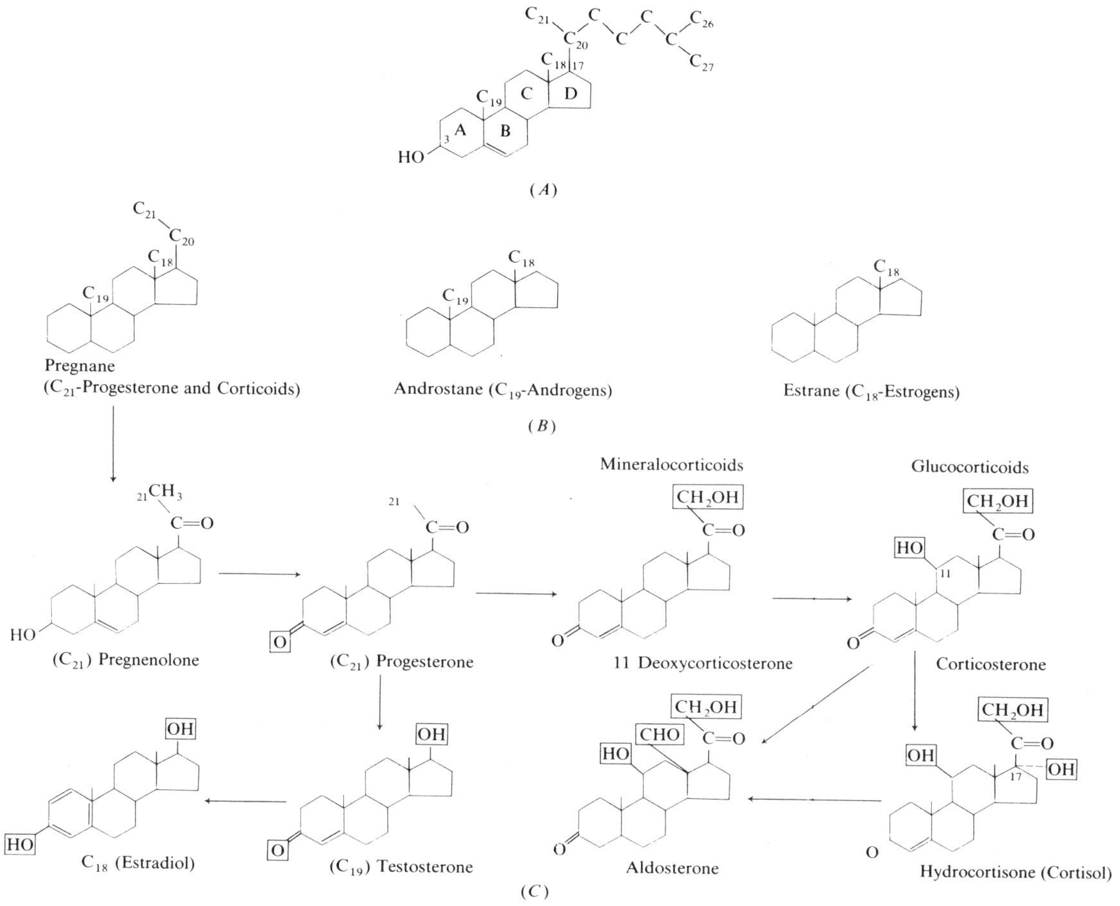

FIGURE 2-8
Structure of steroid compounds. The precursor, cholesterol, is shown in *A*. The three basic types of steroid hormone structures are shown in *B*. In *C* the sequence of synthesis of steroid hormones is shown. Deoxycorticosterone and aldosterone both stimulate increased sodium reabsorption by the kidney.

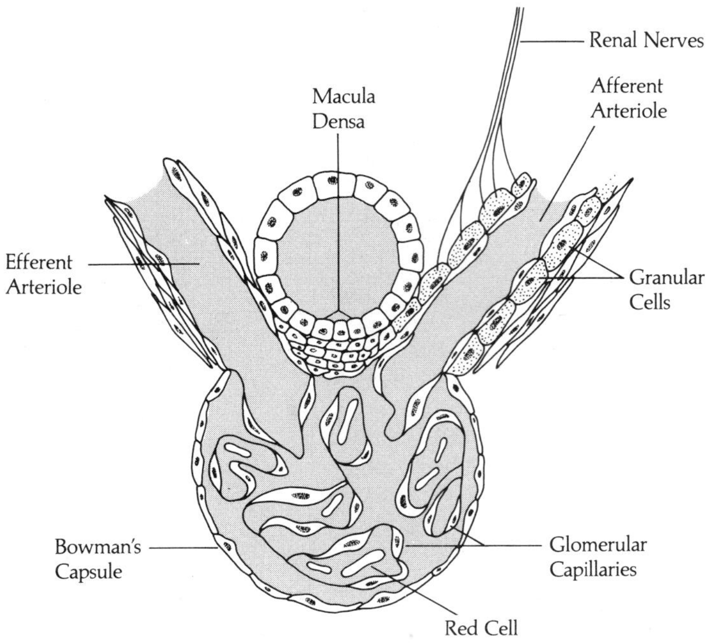

FIGURE 2-9
A diagram of the juxtaglomerular apparatus. Renin is thought to be stored in the granular cells (as granules) and released into the blood when the renal nerves are stimulated or when there is decreased renal blood pressure. *Source:* Modified from *Renal Physiology*, 2nd Ed., by A. J. Vander. Copyright © 1980 by McGraw-Hill Book Company. (This was also modified from James O. Davis, *American Journal of Medicine*, 55:333, 1973.) Reprinted by permission.

renin release. Epinephrine and stimulation of renal sympathetic nerves both increase renin secretion. Sympathetic nerves to the kidney would be activated in hypovolemic and cardiogenic shock. Renin secretion is also related to the amount of sodium flowing into the macula densa. The mechanism by which the sodium mass in the macula densa affects renin secretion has not been resolved. Finally, angiotensin itself acts as a negative feedback to renin release; thus, the more angiotensin in the plasma the greater the inhibition of renin release. It should be remembered that renin release depends on a variety of inputs. The actual amount of renin released depends on the balance between stimulation and inhibition of the granular cells.

SUMMARY

Both external and internal exchange processes are involved in maintaining fluid and electrolyte homeostasis. The key principle is as follows: for balance, input = output. Regulation of volume and composition of interstitial fluid is accomplished by means of diffusion of solutes and bulk flow of fluid, as described by the Starling hypothesis. Excess fluid and protein in the interstitial space is removed through lymphatics. Intracellular volume is regulated by osmosis across cell membranes. Regulation of osmolality of body fluids and volume of extracellular fluid is accomplished through complex endocrine feedback control systems, which include the kidneys, the adrenal glands, the hypothalamus, the posterior pituitary, and the cardiovascular system.

Blood Gases and Blood-Gas Transport

CHAPTER 3

Inferences derived from blood-gas analysis become clear only when one has a thorough understanding of the physiology of the transport of oxygen and carbon dioxide between lungs and peripheral tissues (blood-gas transport). The focus of this chapter is on the physical and chemical aspects of blood-gas transport. The discussion is divided into the following categories: first, a brief description of where blood-gas transport fits into the process of respiration; second, an outline of pertinent gas laws; third, a discussion of the physiologic mechanisms involved in blood-gas transport; and finally the interaction between O_2 and CO_2 transport.

THE RELATION OF BLOOD-GAS TRANSPORT TO RESPIRATION

Respiration is defined as the process of gas exchange between living cells and the external environment. There are three major stages in respiration (Figure 3-1). (1) Ventilation, the first stage, is the flow of a mixture of gases called air into and out of the lungs. (2) Transport, the second stage, has several steps: (a) diffusion of gases into and out of the blood in both pulmonary and systemic capillaries, (b) the chemical and physical reactions of CO_2 and O_2 with blood water and its solutes, and (c) the circulation of blood between pulmonary capillaries and systemic tissue capillaries. (3) During cell respiration, the final

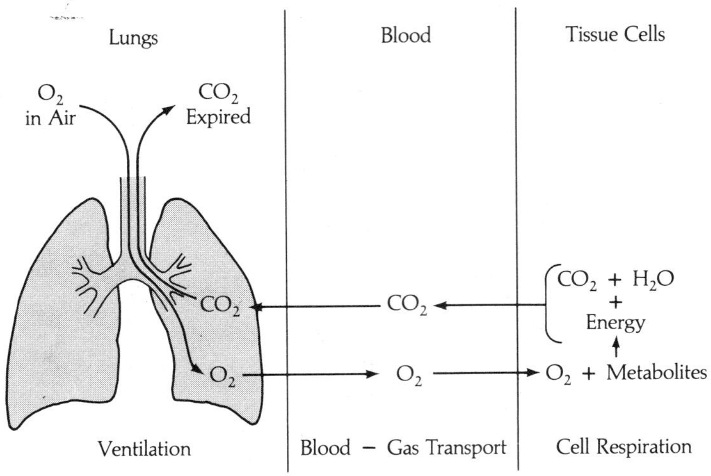

FIGURE 3-1
The three major phases of respiration. The three phases are in series with each other, thus failure in one phase impairs respiration generally. *Source:* "Blood gases and blood-gas transport," by J. L. Keyes, *Heart and Lung*, 1974, 3:945–954. Reprinted by permission of The C. V. Mosby Company.

stage, metabolites are oxidized to obtain energy for life processes, while CO_2 is produced as a waste by-product of metabolism. Steps b and c of the second stage constitute the process of blood-gas transport, the connecting link between ventilation and cell respiration.

GAS LAWS PERTINENT TO THE STUDY OF BLOOD-GAS TRANSPORT

Two of the gas laws, Dalton's law of partial pressures and Henry's law, are essential to this discussion because most of the principles of blood-gas transport relate directly to them.

Dalton's Law of Partial Pressures

Dalton discovered that the total pressure of a given volume of a gas mixture is equal to the sum of the separate or partial pressures that each gas would exert if it alone occupied the entire volume. This relationship is expressed quantitatively in Equation 3-1:

$$P_{Total} = P_1 + P_2 + P_3 + \cdots + P_n \tag{3-1}$$

where P represents pressure, and the subscripts represent different gases in the mixture. For air the equation becomes

$$P_B = P_{N_2} + P_{O_2} + P_{H_2O} + P_{CO_2} \tag{3-2}$$

Table 3-1 Vapor Pressure of Water at Various Temperatures

Temperature, °C	P_{H_2O}, torr
35	42.18
36	44.56
37	47.07
38	49.69
39	52.44
40	55.32

Source: "Blood gases and blood-gas transport," by J. L. Keyes, *Heart and Lung*, 1974, 3:945–954. Reprinted by permission of The C. V. Mosby Company.

where P_B is total (or more commonly, barometric) pressure, P_{N_2}, P_{O_2}, and P_{CO_2} indicate the partial pressures of nitrogen, oxygen, and carbon dioxide, respectively, and P_{H_2O} is the vapor pressure of water.*

The partial pressure of a gas can be calculated if the total pressure and fractional concentration of that gas are known. For example, the partial pressure of oxygen in dry air can be calculated from Equation 3-3:

$$P_{O_2} = F_{O_2} \times P_B \qquad (3\text{-}3)$$

where F_{O_2} is the fractional concentration ($\%O_2 \div 100$) of oxygen in dry air. In dry air oxygen is 20.95% of the total gas mixture, therefore,

$$F_{O_2} = 20.95\% \div 100 = 0.2095 \qquad (3\text{-}4)$$

If P_B is 760 torr, then

$$P_{O_2} = (0.2095)(760 \text{ torr}) = 159 \text{ torr} \qquad (3\text{-}5)$$

Water Vapor Pressure The terms vapor and gas are not strictly synonymous. A vapor is the gaseous state of a substance which also exists simultaneously in a liquid or solid state. Hence, one speaks of water as a vapor, since water exists in both gaseous and liquid states at either body or room temperature. Oxygen, on the other hand, is defined as a gas because it has neither a solid nor liquid phase under the same conditions.

Vapor pressure is the pressure exerted by the vapor phase of the substance. For any pure substance, the vapor pressure depends only on the ambient temperature. It is independent of other factors such as volume and total pressure. Water vapor pressure increases as temperature increases and decreases as temperature decreases (Table 3-1).

---

*CO_2 constitutes less than 1% of the total gases in air.

Table 3-2 Partial Pressures of Gases in Alveolar Air

Gas	Pressure, torr
N_2	564
CO_2	40
H_2O	47
O_2	109
Total pressure	760

Source: "Blood gases and blood-gas transport," by J. L. Keyes, *Heart and Lung*, 1974, 3:945–954. Reprinted by permission of The C. V. Mosby Company.

Water molecules in the vapor phase behave like any other gas; they exert a pressure proportional to their concentration. In effect, then, water molecules in the vapor phase "dilute" the other gases in a mixture so that in a "wet" gas their partial pressures will be reduced in relation to the pressures the other gases would have if the mixture were "dry," provided the total pressure is constant. The following example will help clarify this point. In Equation 3-5 the P_{O_2} in a sample of dry air was 159 torr. Now, if that air were saturated with water at 37°C at one atmosphere pressure, what is the new P_{O_2}? Since we wish to determine the pressure of the inspired air that is oxygen and not oxygen saturated with water, we must subtract the P_{H_2O} from P_B:

$$P_{O_2} = (P_B - P_{H_2O})F_{O_2} \tag{3-6}$$

From Table 3-1, at 37°C $P_{H_2O} = 47$ torr. Therefore,

$$P_{O_2} = (760 \text{ torr} - 47 \text{ torr})(0.2095) \tag{3-7}$$

$$P_{O_2} = (713 \text{ torr})(0.2095) = 149 \text{ torr} \tag{3-8}$$

Thus, inspired air, having gathered water in its passage through the upper airways, has a P_{O_2} 10 torr less than that found in dry air. The P_{O_2} in the alveoli, however, is approximately 100 torr, not 149 torr. The reason for the reduction in P_{O_2} is that inspired air reaching the alveoli is diluted in the alveolar air, which has a water vapor pressure of 47 torr and a P_{CO_2} of 40 torr (Table 3-2).

Henry's Law

Henry found that the amount of gas that will dissolve in a liquid at a specific temperature is directly proportional to the partial pressure of that gas in the gas phase in equilibrium with the liquid. *This law applies only to that fraction of the gas that is physically dissolved in the liquid and not to the fraction of gas that is combined chemically with either the liquid or a solute within the liquid.* Both O_2

Table 3-3 Solubility Coefficient of O_2 and CO_2 in Plasma at 37°C

Gas	Solubility Coefficient	
	Vol % / torr	mmol / L · torr
O_2	0.003	0.0013
CO_2	0.067	0.0301

Source: "Blood gases and blood-gas transport," by J. L. Keyes, *Heart and Lung*, 1974, 3:945–954. Reprinted by permission of The C. V. Mosby Company.

and CO_2 dissolve physically in the water of the blood. However, CO_2 also reacts chemically with water and both CO_2 and O_2 react chemically with hemoglobin. Henry's law applies only to the physically dissolved gas that exerts a partial pressure in the liquid and not the gas that is chemically combined.

When the partial pressure of a gas in a liquid phase is equal to the partial pressure of the same gas in the gas phase, the gas in the two phases is said to be in equilibrium. That is, for every molecule of gas dissolving in the liquid at equilibrium, another molecule of the same gas leaves the liquid. Any change in partial pressure of the gas in either phase will disrupt the equilibrium and cause a corresponding change in the opposite phase until equilibrium is reestablished.

The amount or volume of gas that dissolves in a given volume of liquid depends on the solubility of that gas in the liquid as well as the partial pressure. According to Henry's law the concentration of gas found at equilibrium in a liquid is

$$\frac{\text{Amount of dissolved gas}}{\text{Volume of liquid}} = S \cdot P \qquad (3\text{-}9)$$

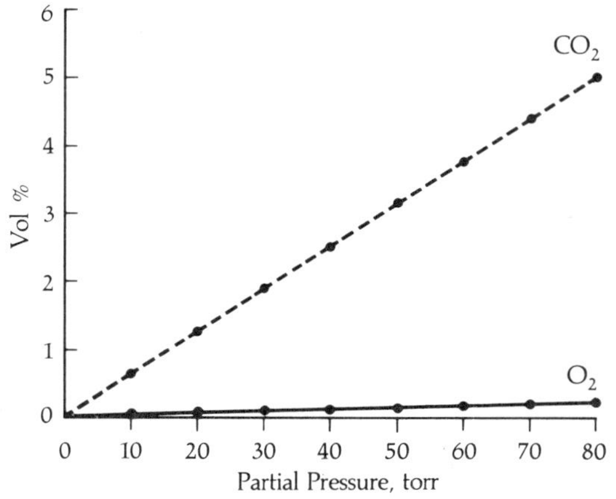

FIGURE 3-2
Graphic representation of Henry's law for both O_2 and CO_2 at body temperature. The concentration of physically dissolved gas in volumes percent increases linearly as partial pressure for that gas increases. Carbon dioxide has a much greater solubility than O_2. *Source:* "Blood gases and blood-gas transport," by J. L. Keyes, *Heart and Lung*, 1974, 3:945–954. Reprinted by permission of The C. V. Mosby Company.

where S is the solubility coefficient of the gas and P the partial pressure of the gas. It should be noted that the value of the solubility coefficient depends on the units chosen to express concentration. If the units are volume percent (vol %), that is, milliliters of gas/100 ml liquid, then S will have units of volume percent per torr gas pressure. If the units are moles per liter then S will have units of moles of gas/liter · torr. Solubility of a gas also varies inversely with temperature; hence, water at higher temperatures will have less physically dissolved gas than water at cooler temperatures. The solubility coefficients of O_2 and CO_2 in plasma are shown in Table 3-3. The relationship between partial pressure and concentration of O_2 and CO_2 is shown in Figure 3-2.

MECHANISMS OF OXYGEN TRANSPORT IN BLOOD

The transport of both oxygen and carbon dioxide in blood is facilitated by hemoglobin. The concentration of hemoglobin in red blood cells is about 35 g in each 100 ml of red blood cells. Thus, in a normal person with a hematocrit of 40%, the hemoglobin concentration is 14 g/dl (dl is the abbreviation for deciliter or 100 ml). Each hemoglobin molecule can combine chemically with four molecules of O_2, and the process is reversible. The reaction is shown schematically in Figure 3-3. The oxygen physically dissolved in the red cell water is in equilibrium with oxygen dissolved in the plasma water. Therefore, at equilibrium the P_{O_2} of plasma water equals the P_{O_2} of water in the red blood cell. The physically dissolved oxygen is also in equilibrium with oxyhemoglobin (HbO_2) in the red cell.

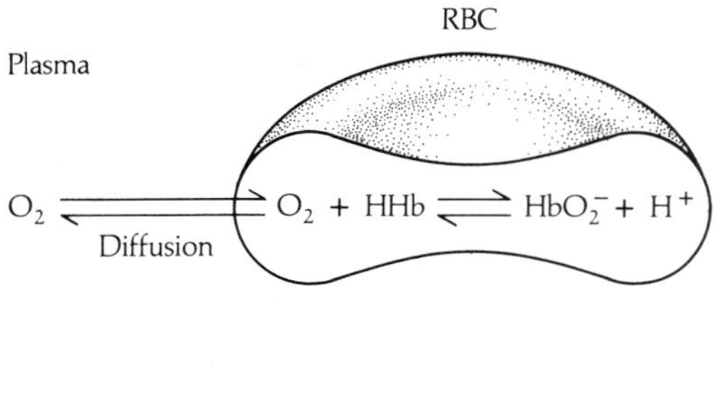

FIGURE 3-3
Oxygen transport in blood. Oxygen physically dissolved in plasma is in equilibrium with the O_2 physically dissolved in red blood cell (RBC) water. Therefore, the P_{O_2} of plasma equals the P_{O_2} of RBC water. The amount of oxygen combined with hemoglobin (Hb) is directly proportional to the P_{O_2} of RBC water. The symbol for saturated (oxyhemoglobin) is HbO_2^- and HHb is the symbol for unsaturated hemoglobin. The equilibrium shifts to the right in the lungs and to the left in the systemic capillaries. *Source:* "Blood gases and blood-gas transport," by J. L. Keyes, *Heart and Lung,* 1974, 3:945–954. Reprinted by permission of The C. V. Mosby Company.

Oxygen Transport in Blood Oxygen is carried in two forms in blood: (1) in physical solution as oxygen dissolved in plasma and red cell water and (2) in reversible chemical combination with hemoglobin. The amount of O_2 transported in *both* forms is directly related to the P_{O_2}.

Physically dissolved oxygen comprises a little over 1% of the total amount of O_2 carried in arterial blood. At 37°C plasma carries 0.003 vol % O_2 per torr P_{O_2}. Thus, at a P_{O_2} of 100 torr, plasma contains 0.3 vol % O_2 (Figure 3-2). This amount of physically dissolved O_2 is not sufficient to sustain life at normal values of cardiac output. Even if we breathe pure O_2, and alveolar P_{CO_2} is constant, the alveolar P_{O_2} would be only 673 torr.* The total O_2 physically dissolved would be approximately 2 vol % (0.003 vol %/torr P_{O_2} × 673 torr = 2.02 vol %). Again, this volume is insufficient to meet body needs at normal cardiac output even at rest. Some kind of carrier or transporter is required. Hemoglobin carries oxygen in a chemically bound form so that the bound O_2 does not contribute to the P_{O_2}.

In effect, when O_2 combines with hemoglobin, the O_2 molecule becomes part of hemoglobin and cannot act independently to exert a partial pressure. About 98% of all O_2 delivered to systemic tissues is transported in chemical combination with hemoglobin. One gram of hemoglobin can carry 1.39 ml O_2. Therefore, 100 ml of blood containing hemoglobin in a concentration of 15 g/dl can carry 20.9 ml of O_2 combined with hemoglobin when the hemoglobin is completely (100%) saturated (Equation 3-10).

$$\text{Vol \% } O_2 = \frac{1.39 \text{ ml } O_2}{\text{g Hb}} \times \frac{15 \text{ g Hb}}{100 \text{ ml blood}} = 20.9 \text{ vol \%} \qquad (3\text{-}10)$$

The amount of oxygen combined with hemoglobin depends on the partial pressure of oxygen. In Figure 3-4 note that at a normal arterial P_{O_2} of 100 torr, whole blood contains 63 times more oxygen per 100 ml than does plasma. The sigmoid shape of the curve has important physiological consequences. At partial pressures greater than 80 torr the curve is nearly flat. This flatness allows for oxygenation of most of the hemoglobin (greater than 90% at 80 torr) in spite of normal variations in alveolar and/or arterial P_{O_2}. The steeper portion of the curve between 15 and 60 torr ensures that large quantities of oxygen can be unloaded from hemoglobin in response to small changes in P_{O_2} within systemic tissue capillaries.

As arterial blood flows through systemic capillaries, oxygen diffuses from the high P_{O_2} of arterial plasma to the lower P_{O_2} of interstitial fluid. As the P_{O_2} of plasma decreases, the physically dissolved O_2 diffuses from the red cells causing a decrement in the P_{O_2} of red cell water. As a result of the decreased red cell P_{O_2}, oxygen dissociates from oxyhemoglobin to become free oxygen physically dissolved in red cell water and thus is able to diffuse into regions of lower P_{O_2} (see Figures 3-3 and 3-8).

It is apparent that the degree of oxygen dissociation and association with hemoglobin is a direct function of the P_{O_2}. After the blood passes through the systemic capillaries, the hemoglobin is not competely desaturated. Rather, there is a large reserve of oxygen available to meet increasing tissue needs. At rest at normal hemoglobin concentrations there is more than 14 vol % O_2 in

---

*P_B = 760 torr and P_{H_2O} = 47 torr. Therefore, $P_{O_2} = P_B - P_{H_2O} - P_{CO_2}$. The P_{CO_2} of alveolar air is 40 torr. The assumption is made that P_{N_2} = 0, which would be true only after several hours of breathing pure O_2.

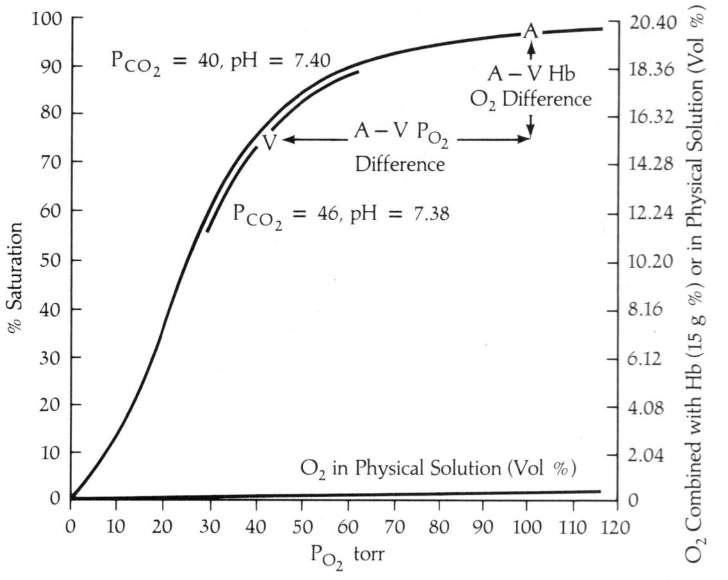

FIGURE 3-4
A comparison of the amount of O_2 carried in whole blood in volumes percent as a function of P_{O_2}. The symbol, *A*, labels the arterial whole blood curve and *V*, the mixed venous whole blood curve. The venous curve is shifted slightly to the right and below the arterial curve. The volume percent (*right*) ordinate is labeled for whole blood containing a hemoglobin concentration of 15 g / dl. The P_{CO_2} = 40, pH = 7.4 label refers to the arterial curve whereas the P_{CO_2} = 46, pH = 7.38 refers to the mixed venous curve. The symbol, *A* – *V*, represents the arterial venous difference. The left ordinate indicates percent saturation of the hemoglobin. *Source:* "Blood gases and blood-gas transport," by J. L. Keyes, *Heart and Lung*, 1974, 3:945–954. Reprinted by permission of The C. V. Mosby Company.

mixed venous blood, most of which is still combined with hemoglobin (Figure 3-4).

The Effect of Hemoglobin Concentration on Oxygen Transport The more hemoglobin that is in blood, the more oxygen can be carried by that blood at a given P_{O_2} (Figure 3-5). If hemoglobin concentration is increased to 20 g/dl (hematocrit approximately 0.6 or 60%), then at a P_{O_2} of 100 torr the content of O_2 in arterial blood would be 26.5 vol %. On the other hand, in anemia when hemoglobin concentration may be as low as 10 g/dl (hematocrit approximately 0.27) at a P_{O_2} of 100 torr the oxygen content would be 13 vol %. In each case, as shown in Figure 3-5, the hemoglobin is 97% saturated at the same P_{O_2}, but the content or concentration of oxygen varies twofold over the hemoglobin concentration range. Therefore, not only is it important for an individual to saturate his hemoglobin, it is equally important to have enough hemoglobin to carry a sufficient supply of O_2. For example, in strenuous exercise oxygen consumption in muscle tissues can increase fourfold over that at rest. During exercise, the individual with anemia may have difficulty meeting tissue oxygen requirements because this individual may lack sufficient hemoglobin to carry the volume of O_2 in his or her blood required by the increased demand. Even more important, in anemia the myocardium may not receive a sufficient supply of oxygen to meet the increased demands of exercise.

From the preceding discussion it is apparent that both P_{O_2} and total concentration of oxygen content are important in assessing an individual's status in terms of blood-gas transport. However, it must be remembered that even with normal hemoglobin concentrations and normal arterial P_{O_2}, the

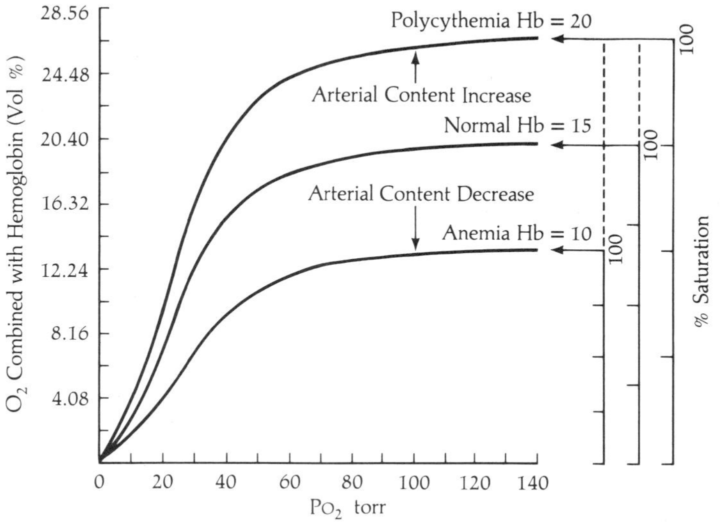

FIGURE 3-5 Oxygen saturation of whole blood as a function of hemoglobin concentration. Note that percent saturation of hemoglobin is independent of hemoglobin concentration. Hemoglobin concentrations are given in grams per deciliter of blood. To determine the concentration (volume %) of O_2 carried in a patient's blood, multiply the hemoglobin concentration by 1.39 ml O_2 per gram of hemoglobin. Then multiply that product by the percent saturation/100. *Source:* "Blood gases and blood-gas transport," by J. L. Keyes, *Heart and Lung*, 1974, 3:945–954. Reprinted by permission of The C. V. Mosby Company.

systemic tissues may not receive an adequate supply of oxygen if the cardiovascular system is malfunctioning. Tissue cells, even at rest, require a minimum amount of O_2 to maintain life. This O_2 is delivered by the flow of blood. Any event that reduces blood flow will also reduce the delivery of oxygen. The end result may be tissue hypoxia. For example, following a myocardial infarction a person may have a markedly reduced cardiac output while his or her arterial P_{O_2} and hemoglobin concentration may be quite normal. Nonetheless, the systemic tissues may be hypoxic because of inadequate oxygen delivery. *Therefore, it should be remembered that blood-gas transport of oxygen involves both the physical and chemical reactions of O_2 in blood and the delivery of that oxygen in blood to the systemic tissues.*

CARBON DIOXIDE TRANSPORT IN BLOOD

At rest the body produces CO_2 at a rate of approximately 200 ml/minute. This carbon dioxide is excreted from the body through the lungs. The carbon dioxide must be transported from the systemic tissues to the lungs in blood. Carbon dioxide is carried in blood in four forms: (1) physically dissolved CO_2, (2) carbonic acid, (3) carbamino compounds, and (4) bicarbonate.

Physically Dissolved CO_2 Carbon dioxide is 22 times more soluble than oxygen in water at the same temperature. At a P_{CO_2} of 40 torr and at 37°C the CO_2 content of blood water is about 2.5 vol %. Physically dissolved CO_2 accounts for approximately 8% of the CO_2 transported from tissues to lungs for excretion, that is, the difference between arterial and venous bloods.

Carbonic Acid (H_2CO_3) Carbon dioxide is also transported as carbonic acid (Equation 3-11).

$$CO_2 + H_2O \rightleftharpoons H_2CO_3 \rightleftharpoons H^+ + HCO_3^- \qquad (3\text{-}11)$$

There is one molecule of H_2CO_3 for every 340 molecules of CO_2. Thus, the amount of CO_2 carried in this form accounts for less than 1% of the total CO_2 transported. Because the amount of CO_2 transported as carbonic acid is small, it is usually considered or lumped together with physically dissolved CO_2. It should be noted that carbonic acid is an important component of Reaction 3-11 and cannot be ignored in the reaction. The conversion of CO_2 to carbonic acid occurs in both plasma and red cells. The reaction proceeds very slowly in plasma, but is accelerated enormously in the red cell by the enzyme, carbonic anhydrase.* Because of carbonic anhydrase the reaction shown in Equation 3-11 can essentially proceed to equilibrium in the time the blood is in the capillary.

Carbamino Compounds Carbon dioxide can combine chemically with certain amino groups on proteins in plasma and hemoglobin in red cells. These carbamino compounds account for about 11% of the CO_2 transported. The amount of CO_2 transported combined with plasma proteins is negligible compared to that combined with hemoglobin. Thus, carbamino hemoglobin represents the vast majority of the 11% of CO_2 transported in this form. The reaction of CO_2 and hemoglobin is shown in Equation 3-12:

$$Hb - NH_2 + CO_2 \rightleftharpoons Hb - NHCO_2^- + H^+ \qquad (3\text{-}12)$$

The reaction proceeds quickly and requires no enzyme to catalyze the reaction. Deoxygenated (reduced) hemoglobin combines with more CO_2 than oxyhemoglobin. Oxygenation of reduced hemoglobin reverses Reaction 3-13 whereas reduction of the hemoglobin promotes combination of CO_2 with hemoglobin (see Figure 3-11). It should be noted that CO_2 does not combine at the same site as oxygen. Carbon dioxide combines with the amino ($-NH_2$) groups of

*The reaction actually couples CO_2 to the OH^- as follows: $CO_2 + OH^- \rightarrow HCO_3^-$. The hydroxyl ($OH^-$) ion comes from dissociation of H_2O: $H_2O \rightleftharpoons H^+ + OH^-$. However, for convenience the reaction will be shown in the more conventional reaction, Equation 3-12.

the protein globin components, whereas O_2 combines with the Fe^{++} ion, which is part of the heme component.

Bicarbonate The majority of CO_2 transported from systemic tissues to lungs is in the form of bicarbonate (Equation 3-11). Approximately 81% of the CO_2 to be excreted is carried in blood as HCO_3^-. Nearly all the HCO_3^- transported as a result of addition of CO_2 is produced in the red cell because the red cell contains carbonic anhydrase. Plasma contains virtually none of this enzyme, hence the reaction though occurring in plasma is very slow and accounts for very little of the HCO_3^- found in the plasma. As HCO_3^- is produced, its concentration increases in red cell water and it diffuses from the red cell into plasma. The hydrogen ions formed in the reaction tend to be retained and buffered by hemoglobin. In exchange for the HCO_3^-, chloride diffuses from plasma into the red cells. This exchange is called the chloride shift (Figure 3-6). Thus, most of the HCO_3^- produced by adding CO_2 to blood is transported in plasma even though the major fraction of the HCO_3^- is synthesized in the red blood cell. The steps shown in Figure 3-6 are reversed when blood flows through the pulmonary capillaries in the lungs (Figure 3-7).

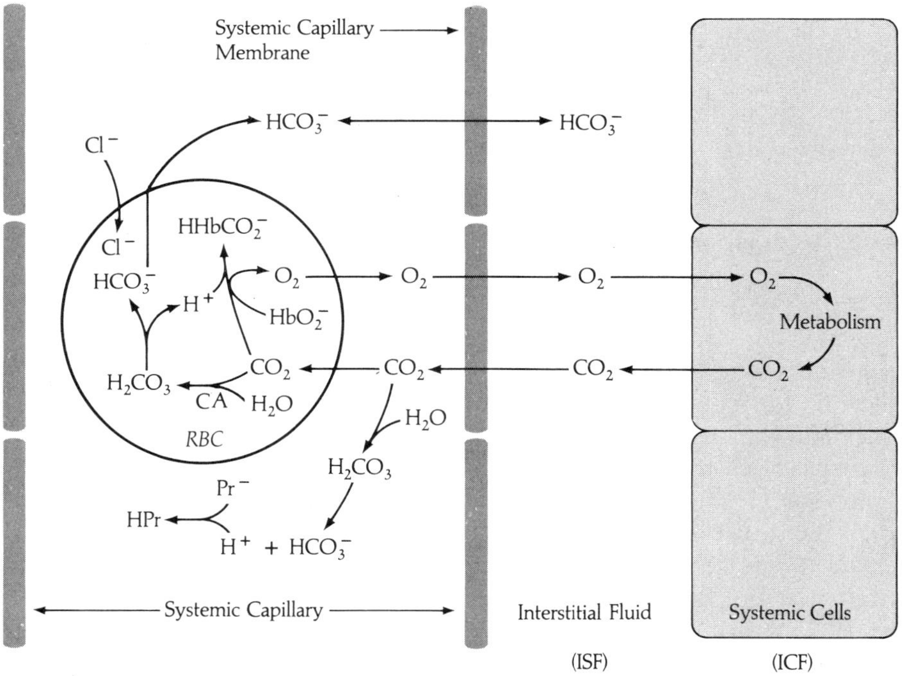

FIGURE 3-6 A summary of gas exchange in systemic tissue capillaries. Hb is hemoglobin, CA is carbonic anhydrase, Pr^- is plasma protein, and $HHbCO_2^-$ is carbaminohemoglobin.

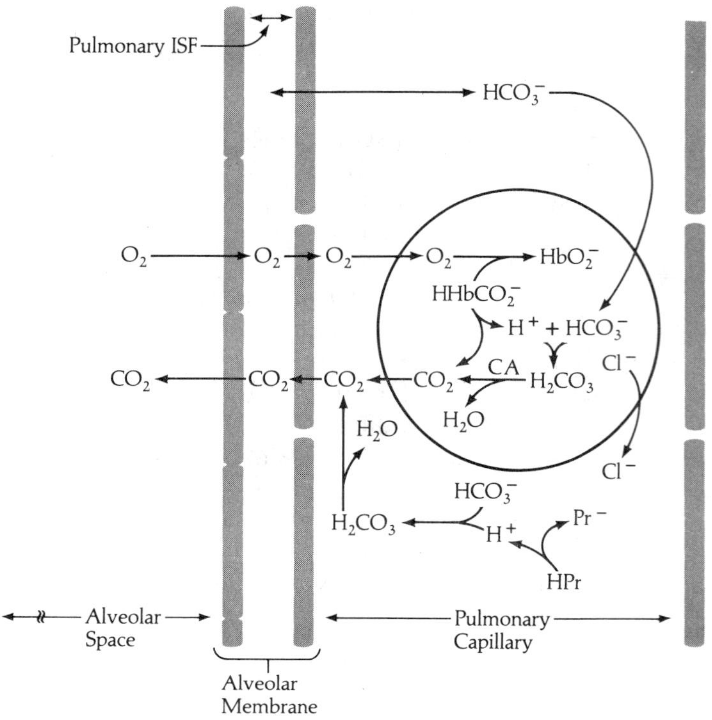

FIGURE 3-7
Summary of gas exchange in pulmonary capillaries. The symbols used are the same as those shown in Figure 3-6.

FACTORS AFFECTING O_2 AND CO_2 TRANSPORT

Oxygen Transport and P_{50}

The hemoglobin dissociation curve can be used to show changes in the affinity of hemoglobin for O_2. When studying affinity, the percentage saturation of hemoglobin with O_2 is used for the ordinate axis instead of concentration in order to eliminate the effect of hemoglobin concentration as a variable. The position of the dissociation curve can be described in terms of the P_{50}, that is, the partial pressure of O_2 at which the hemoglobin is 50% saturated. The P_{50} of normal adult hemoglobin is 26 torr (Figure 3-8). If the P_{50} is increased (shifted to the right), the affinity of hemoglobin for O_2 is decreased. When the P_{50} is decreased (shifted to the left) hemoglobin has a higher affinity for O_2. Five factors are of considerable physiologic importance in altering the P_{50} of hemoglobin: (1) blood pH, (2) P_{CO_2}, (3) temperature, (4) 2,3-diphosphoglycerate (2,3-DPG), and (5) carbon monoxide.

The pH of blood has long been known to alter the affinity of hemoglobin for O_2. Increased pH (alkalemia) causes a reduction in the P_{50} (increased affinity) whereas a decrease in pH increases the P_{50} (decreased affinity) (Figure 3-8).

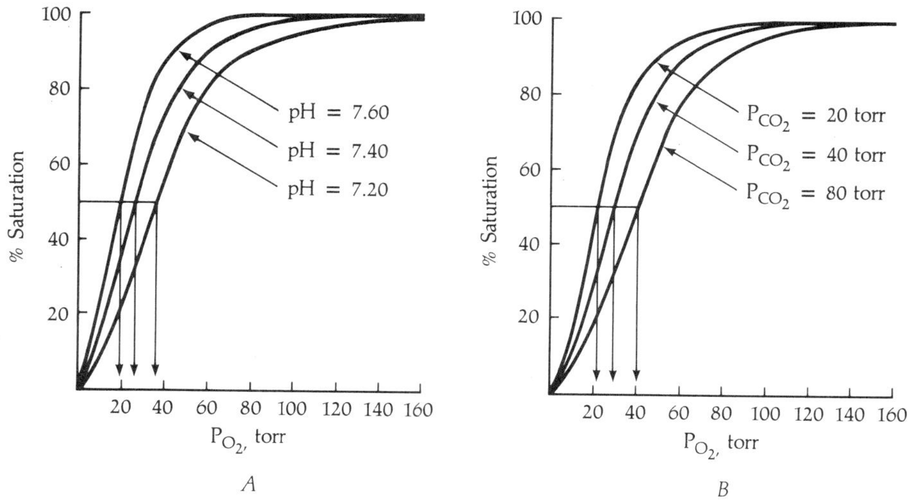

FIGURE 3-8 The Bohr effect. In *A* the shift in P_{50} is shown for change in pH whereas in *B* the Bohr effect is shown for changes in P_{CO_2}. The vertical arrows in both figures point to the P_{50} for each curve. *Source:* Figure 3-8a, "Blood gases and blood-gas transport," by J. L. Keyes, *Heart and Lung*, 1974, 3:945–954. Reprinted by permission of The C. V. Mosby Company.

Temperature has a marked effect on affinity as shown in Figure 3-9. In tissues where metabolic rate is high, both temperature and $[H^+]$ are increased. Both of these effects tend to increase the P_{50} and promote release of O_2 from hemoglobin. Temperature of blood in the pulmonary circuit is reduced slightly compared to core temperature because of ventilation, and in the pulmonary circuit pH tends to rise slightly because of loss of CO_2. Both of these latter effects reduce P_{50} and promote "loading" of O_2 on hemoglobin.

FIGURE 3-9
The oxyhemoglobin dissociation curve at three different temperatures. Note the shift in the P_{50} to the left (*vertical arrows*) as temperature decreases.

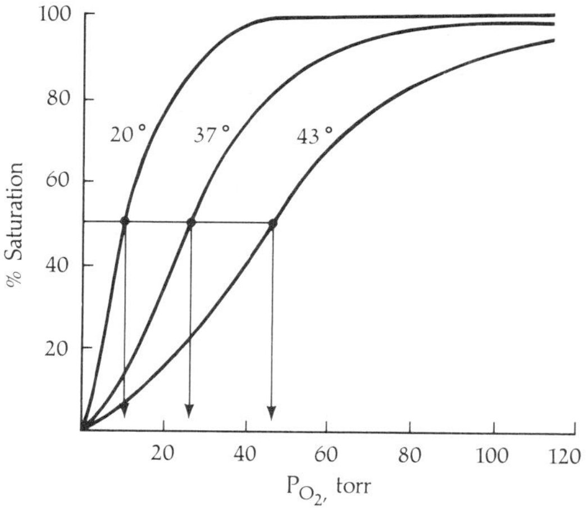

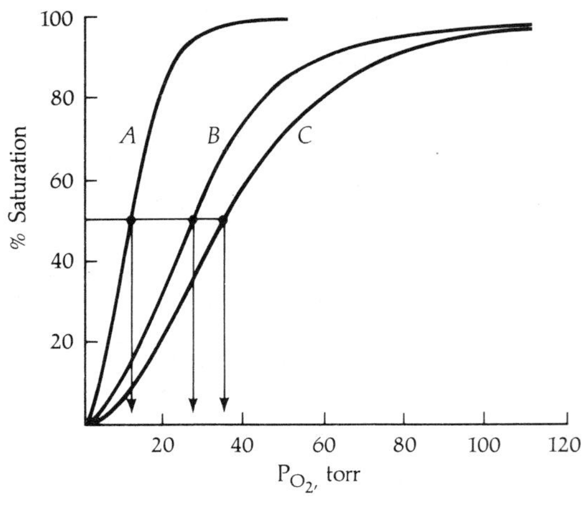

FIGURE 3-10
The effect of 2,3-DPG on the position of the oxyhemoglobin dissociation curve. (Curve *A*) no 2,3-DPG; (Curve *B*) normal blood; (Curve *C*) higher than normal concentration of 2,3-DPG. Note the shift in P_{50} to the left when 2,3-DPG is absent.

Several substances bind with hemoglobin and help to reduce the very high affinity of hemoglobin for oxygen. These substances (or ligands) include H^+, CO_2, and 2, 3-DPG. The effects of H^+ are shown in Figure 3-8. The effects of CO_2 are discussed below (Bohr effect). The ligand 2, 3-DPG is one of many organic polyphosphate anions that bind to hemoglobin. When 2, 3-DPG binds with hemoglobin, the P_{50} increases, hence the affinity of O_2 is reduced (Figure 3-10). Red blood cells contain more 2, 3-DPG than other cells of the body. The ratio of 2, 3-DPG molecules to hemoglobin molecules is almost 1:1. Molecules of 2, 3-DPG are an intermediate in the glycolysis pathway and the concentration of the ligand is closely regulated in the red cell. When blood is stored in a refrigerated container as it is at a blood bank, the concentration of 2, 3-DPG decreases. As a result, the affinity of hemoglobin for O_2 increases. Hence, although transfusions of stored blood increase blood volume, they may not improve O_2 delivery to tissue as much as expected in the first couple of hours after transfusion. After a few hours, the concentration of 2, 3-DPG will be restored to normal.

Carbon monoxide combines with hemoglobin at the same site O_2 combines. However, hemoglobin has a 210-fold greater affinity for carbon monoxide (CO) than it does for O_2. When P_{CO} is 0.12 torr, 50% of the hemoglobin is saturated with CO. In addition, CO shifts the oxyhemoglobin curve to the left; thus, with CO poisoning, O_2 delivery at systemic tissues will be further reduced because of the decreased P_{50}.

The Bohr and Haldane Effects

There are two types of interaction between CO_2 transport and O_2 transport: (1) the effect of CO_2 on oxygen transport by hemoglobin and (2) the effect of oxygen on the transport of CO_2. These interactions affect the chemical reactions of the gases, not the diffusion of physically dissolved gas.

FIGURE 3-11
The Haldane effect. CO_2 concentration curves are shown at three different values of P_{O_2}, 0, 50, and 100 torr. P_{O_2} has no effect on physically dissolved CO_2. *Source:* "Blood gases and blood-gas transport," by J. L. Keyes, *Heart and Lung*, 1974, 3:945–954. Reprinted by permission of The C. V. Mosby Company.

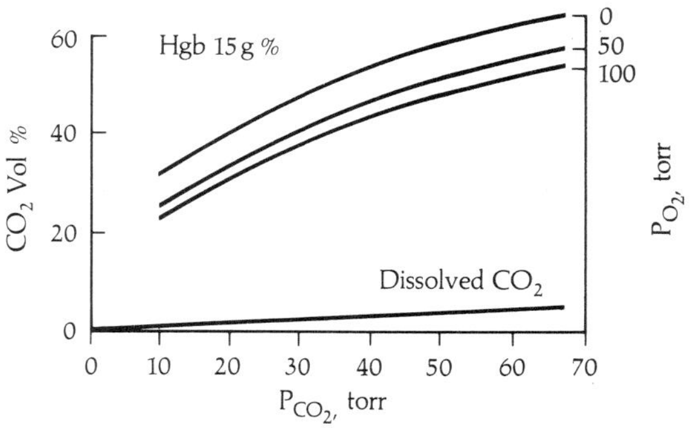

Carbon dioxide acts similarly to other ligands and shifts the P_{50} to the right, thereby reducing the affinity of hemoglobin for oxygen. There are two components to the effect of added CO_2 to blood. First, CO_2 binds with hemoglobin to form carbaminohemoglobin. Second, CO_2 is hydrated to form carbonic acid, which then dissociates, yielding H^+ and HCO_3^- ions. The H^+ also combines with amino groups on hemoglobin and thereby acts as a ligand. Both CO_2 and H^+ shift the P_{50} to the right. The combined action of CO_2 and H^+ on the affinity of hemoglobin for oxygen is called the *Bohr effect*. This shift in the dissociation curve occurs when blood flows through systemic capillaries. Consequently, the addition of CO_2 and H^+ to hemoglobin increases the amount of oxygen given up in systemic capillaries to the tissues.

When oxygen is added to blood, the total CO_2 content of blood is reduced (Figure 3-11). This interaction is called the *Haldane effect*. There are two components to the Haldane effect. If O_2 is added to blood at constant P_{CO_2}, there is a decrease in the concentration of carbamino hemoglobin because of the decreased affinity of hemoglobin for CO_2. The decreased affinity for CO_2 accounts for approximately 70% of the Haldane effect. The remaining 30% is due to the release of hydrogen ions from hemoglobin as it becomes oxygenated. The hydrogen ions combine with HCO_3^- to form H_2CO_3, which in turn dehydrates to CO_2 and H_2O (Equation 3-12). At constant P_{CO_2} the extra CO_2 formed from dehydration of H_2CO_3 is removed from solution. Figure 3-12 compares the Bohr and Haldane effects.

SUMMARY

Oxygen and carbon dioxide dissolve in blood water and also combine chemically with constituents of blood. The chemical reactions are reversible, so both O_2 and CO_2 are easily released from chemical combination with these constituents. The total amount of a given gas in solution is equal to the sum of physically dissolved gas plus that volume chemically and reversibly combined

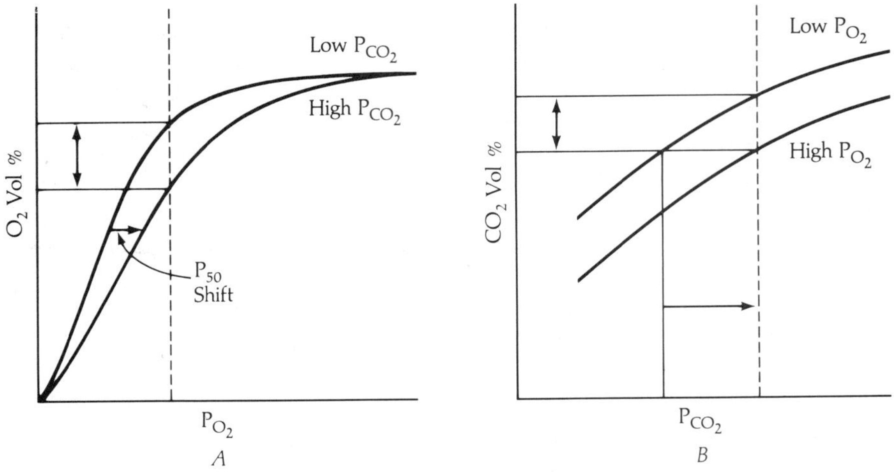

FIGURE 3-12 A comparison of the Bohr (*A*) and Haldane (*B*) effects. When CO_2 is added, (*A*) the curve shifts to the right increasing the P_{50}. The vertical arrow shows the change in oxygen content that would occur at the constant PO_2 indicated by the long vertical dashed line. The Bohr effect is most pronounced in the "middle" ranges of P_{O_2}, that is, from approximately 10 to 80 torr. When hemoglobin is fully saturated or completely desaturated, the Bohr effect cannot be observed. In (*B*), the vertical arrow shows the change in CO_2 content that would occur when O_2 is added to reduced hemoglobin at a constant P_{CO_2}. The horizontal arrow shows the change in P_{CO_2} that would be required to maintain the CO_2 content at a constant value when P_{O_2} increases. *Source:* "Blood gases and blood-gas transport," by J. L. Keyes, *Heart and Lung*, 1974, 3:945–954. Reprinted by permission of The C. V. Mosby Company.

with other substances. The total concentration of both O_2 and CO_2 in blood is much greater than the concentration of physically dissolved O_2 and CO_2.

The partial pressure of a gas in a solution is proportional *only* to the amount of gas physically dissolved and not to total concentration of the gas.* The rates of diffusion, chemical reactions, and physiologic properties of gases depend on the partial pressures of those gases rather than their total concentration in solution. For example, the concentration of physically dissolved O_2, which is proportional to P_{O_2}, affects the rate of diffusion into and out of capillaries and red cells. Furthermore, the P_{O_2} is one of the factors that determine the amount of and rate at which O_2 combines with hemoglobin. Similarly, the P_{CO_2} in blood determines the rate of reactions involving CO_2 and has direct effects on respiration and circulation of blood. On the other hand, the total concentration of O_2 and CO_2 determines the amounts of these gases available for reactions. Thus, both partial pressure and total concentration of a gas are important in respiration.

*Total concentraton is equal to the amount of gas physically dissolved *plus* that chemically combined per unit volume.

Regulation of P_{CO_2} and Bicarbonate Concentration of Plasma

CHAPTER 4

In health, the plasma P_{CO_2} and HCO_3^- concentration are regulated within narrow limits. The processes involved in this regulation play central roles in maintaining acid-base homeostasis. These processes are described briefly in this chapter to provide a framework for understanding acid-base regulation. More detailed discussions of each of these topics may be found in texts of respiratory and renal physiology included in the references for this chapter.

VENTILATION

Breathing and ventilation are not synonymous. Breathing is alternately inspiring and expiring air into and out of the lungs. Ventilation is the rate at which a volume of air moves into or out of the lungs. The units of breathing are breaths per minute whereas ventilation is measured by volume per unit time, usually liters per minute. Ventilation may be determined by measuring the volume of air expired in liters and dividing the result by the time in minutes that it took to obtain that volume. The ventilation determined in this fashion is called the *expired minute volume*, $\dot{V}_E$. The symbol V with a dot over it represents the volume of gas per unit of time or flow and E refers to *expired* volume.* The minute volume may also be calculated from the product of the frequency of

*Any symbol with a dot over it denotes a rate. The symbol V with no dot refers to volume. The symbol E is used to denote that the volume measured is expired not inspired volume.

respiration and tidal volume, as shown in Equation 4-1.

$$\dot{V}_E = V_T \times f \tag{4-1}$$

V_T (no dot) is tidal volume and f is frequency.

Dead Space versus Alveolar Space Minute volume has two components, dead space volume and alveolar volume. The dead space is that volume in the respiratory tract and lungs that does not participate in gas exchange with blood. Dead space includes the volume of airways from the nose to respiratory bronchioles, called *anatomic dead space*, and the volume of gas in alveoli that does not equilibrate with blood, the *alveolar dead space*. The sum of alveolar and anatomic dead space is called total or effective dead space and is also referred to as physiologic dead space. In healthy adults the volume of the dead space is about 150 ml.* At the end of inspiration, the anatomic dead space contains air that is warmed to body temperature and saturated with water vapor but does not exchange O_2 and CO_2 with blood. At the end of expiration that same space contains air that has come from alveoli and has a composition essentially the same as that found in the alveoli.

The alveolar space is where gas exchanges with blood. The alveolar space includes the volume encompassed by the respiratory bronchioles, alveolar ducts, and alveolar sacs or alveoli. The volume of the alveolar space varies considerably during each respiratory cycle at rest and can be varied even more during exercise depending on the magnitude of the tidal volume.

Dead Space Ventilation versus Alveolar Ventilation When we breathe, both dead space and alveolar space are ventilated. The total ventilation or minute volume is the sum of the ventilation of each space as shown in Equation 4-2:

$$\dot{V}_E = \dot{V}_D + \dot{V}_A \tag{4-2}$$

By definition, dead space ventilation ($\dot{V}_D$) does not contribute to gas exchange with blood; only alveolar ventilation ($\dot{V}_A$) is, by definition, the flow of inspired air that participates in gas exchange with blood. In order to determine the value of alveolar ventilation, the dead space ventilation must be measured or calculated. Dead space ventilation is the product of dead space volume (V_D) and frequency of respiration, that is,

$$\dot{V}_D = V_D \times f \tag{4-3}$$

Once $\dot{V}_D$ has been calculated and $\dot{V}_E$ measured, $\dot{V}_A$ may be calculated from Equation 4-2.

*In normal people virtually all of the 150-ml dead space volume is anatomic, that is, the alveolar dead space is very small. In disease, alveolar dead space can increase significantly.

Alveolar ventilation is defined as the volume of *fresh* air entering the alveolar space per minute. When we inspire, the first 150 ml (approximately) of air entering the alveolar space comes from the dead space. Recall that at the end of the previous expiration the dead space was filled with gas exhaled from alveoli. Therefore, dead space air cannot much alter composition of alveolar gas. Any inspired volume greater than dead space volume will add fresh air to the alveolar space and thereby enrich the P_{O_2} and decrease P_{CO_2} of alveolar gas. The relationship between alveolar ventilation rate and alveolar P_{CO_2} is shown in the following equations.*

In a steady state, CO_2 production ($\dot{V}_{CO_2}$) equals CO_2 excretion ($\dot{V}_E \times F_{E_{CO_2}}$) as shown in Equation 4-4,

$$\dot{V}_{CO_2} = \dot{V}_E \times F_{E_{CO_2}} \tag{4-4}$$

where $F_{E_{CO_2}}$ is the fractional concentration of CO_2 in expired gas. Mixed expired gas contains gas from the dead space and gas from the alveolar space. Recall that at the end of inspiration the dead space contains inspired air that has been warmed and saturated with water vapor. Therefore, in a steady state, Equation 4-4 may be rewritten as follows:

$$\dot{V}_{CO_2} = \dot{V}_D F_{I_{CO_2}} + \dot{V}_A F_{A_{CO_2}} \tag{4-5}$$

where $F_{I_{CO_2}}$ is the fractional concentration of CO_2 in the inspired air and $F_{A_{CO_2}}$ is the fractional concentration of CO_2 in alveolar gas. The CO_2 concentration of inspired air is very low (less than 0.04%); therefore, $F_{I_{CO_2}} = 0$. Hence in a steady state,

$$\dot{V}_{CO_2} = \dot{V}_A F_{A_{CO_2}} \tag{4-6}$$

If both sides of Equation 4-6 are multiplied by ($P_B - 47$) we obtain the relationship between CO_2 production, alveolar ventilation rate, and alveolar P_{CO_2} ($P_{A_{CO_2}}$):

$$(P_B - 47)\dot{V}_{CO_2} = \dot{V}_A \cdot P_{A_{CO_2}} \tag{4-7}^{\dagger}$$

Rearranging Equation 4-7 we obtain

$$P_{A_{CO_2}} = \frac{(P_B - 47)(\dot{V}_{CO_2})}{\dot{V}_A} \tag{4-8}$$

*The derivation is very straightforward. However, some prefer to get the answer first and then look at the pathway. The answer is Equation 4-9.

† $F_{A_{CO_2}}(P_B - 47) = P_{A_{CO_2}}$. As explained in Chapter 3, the partial pressure of a gas is equal to the fractional concentration of the gas multiplied by the total pressure.

Table 4-1 Changes in $\dot{V}_A$ and $\dot{V}_D$ Following Changes in Frequency and Tidal Volume for a Constant $\dot{V}_E$ of 10 L / Minute

f breaths / minute	V_D ml	V_T ml	$\dot{V}_D$ L / minute	$\dot{V}_A$ L / minute
10	150	1000	1.5	8.5
12	150	833	1.8	8.2
14	150	714	2.1	7.9
16	150	625	2.4	7.6
20	150	500	3.0	7.0

In a steady state $\dot{V}_{CO_2}$ is constant and at any given time (P_B − 47) is also constant. Since the product of two constants is another constant, Equation 4-8 may be rewritten

$$P_{A_{CO_2}} = \frac{K}{\dot{V}_A} \tag{4-9}$$

where $K = \dot{V}_{CO_2}(P_B - 47)$. Equation 4-9 tells us that alveolar P_{CO_2} is inversely related to alveolar ventilation rate. For a given rate of CO_2 production (i.e., metabolic rate), the greater the alveolar ventilation rate, the smaller will be the value of $P_{A_{CO_2}}$. Conversely, the smaller the value of $\dot{V}_A$ for a given metabolic rate, the greater will be the $P_{A_{CO_2}}$. Because blood normally equilibrates with alveolar gas, changes in alveolar P_{CO_2} will cause corresponding changes in arterial P_{CO_2}. Therefore, determination of arterial P_{CO_2} provides information on the adequacy of alveolar ventilation. When arterial P_{CO_2} is greater than 40 torr, alveolar ventilation is less than it should be for the prevailing $\dot{V}_{CO_2}$ and the subject is said to be *hypoventilating*. When arterial P_{CO_2} is less than 40 torr the alveolar ventilation rate is greater than necessary for the prevailing $\dot{V}_{CO_2}$ and the subject is said to be *hyperventilating*. Note that the definition of hyper- and hypoventilation is determined in terms of P_{CO_2}, not P_{O_2}.

Factors Altering $\dot{V}_D$ and $\dot{V}_A$ For a given minute volume, the greater the dead space ventilation, the smaller will be the corresponding alveolar ventilation rate and vice versa (Equation 4-2). Dead space ventilation increases with rapid shallow breathing and decreases with slower deeper breathing (Equation 4-3 and Table 4-1). Dead space ventilation increases when there is reduction or maldistribution of blood flow to pulmonary capillaries. For example, during pulmonary hypotension alveolar dead space can increase, especially if the subject is sitting or standing. When one is in an upright position, blood flow is greater in the bases than in the apices of the lungs because of the effect of gravity. With pulmonary hypotension there is a reduced flow to the apical alveoli. Consequently, ventilation of these alveoli is less effective than normal for gas exchange with blood. Maldistribution of blood flow can occur with a pulmonary embolism. Again, ventilation of poorly perfused regions of the lungs leads to increased alveolar dead space ventilation. In other words, when regions of the lung are underperfused in relation to the ventilation they receive, overall alveolar dead space ventilation increases. When a recumbent position is assumed by the patient, alveolar dead space decreases because blood flow is more evenly distributed.

Alveolar ventilation can be increased with deep breathing. In the clinical setting, alveolar ventilation can be increased in mechanically ventilated patients by increasing tidal volume. Increasing frequency of respiration increases both $\dot{V}_A$ and $\dot{V}_D$. When arterial P_{CO_2} is too low in a mechanically ventilated patient, an increase may be effected by increasing tubing dead space and/or decreasing tidal volume. The dead space volume may be increased by lengthening the tubing between the expiratory port and the patient. Table 4-1 shows the relationships between $\dot{V}_A$ and $\dot{V}_D$ for a constant minute volume when frequency and tidal volume change.

REGULATION OF VENTILATION

Traditionally, the topic of regulation of ventilation has been divided into neural regulation and chemical regulation. This division is somewhat arbitrary because ultimately all stimuli for changing ventilation must be integrated by the nervous system. Ventilation of the lungs occurs only when the muscles of respiration contract and relax. Contraction of these skeletal muscles is under direct control of the nervous system. A breathing pattern is established by rhythmic stimulation of these muscles. The stimulus for muscle contraction originates in the medullary respiratory centers.*

The rate of ventilation is not constant, but varies to meet demands for increased oxygen consumption and CO_2 excretion. Ventilation rate changes in response to a variety of stimuli from different sources. Figure 4-1 shows a model of the control system. Each incoming stimulus from various receptors must be interpreted and an integrated response generated from the total sensory input. A multitude of stimuli come from chemoreceptors, pulmonary stretch receptors, baroreceptors, plus many others. The chemoreceptors are of primary concern for acid-base regulation because they detect changes in the pH, P_{CO_2}, and in the case of peripheral chemoreceptors, P_{O_2}.

Central Chemoreceptors The central chemoreceptors are located (in cats and dogs) in the medulla oblongata near the exit of the ninth (glossopharyngeal) and tenth (vagus) cranial nerves. The central chemoreceptors respond to changes in P_{CO_2} and H^+ concentrations of the interstitial fluid of the medulla

*An excellent review of the details of the neurogenesis of breathing is presented by Berger et al., *New England Journal of Medicine*, 297, 1977: 92–97, 138–143, 194–201.

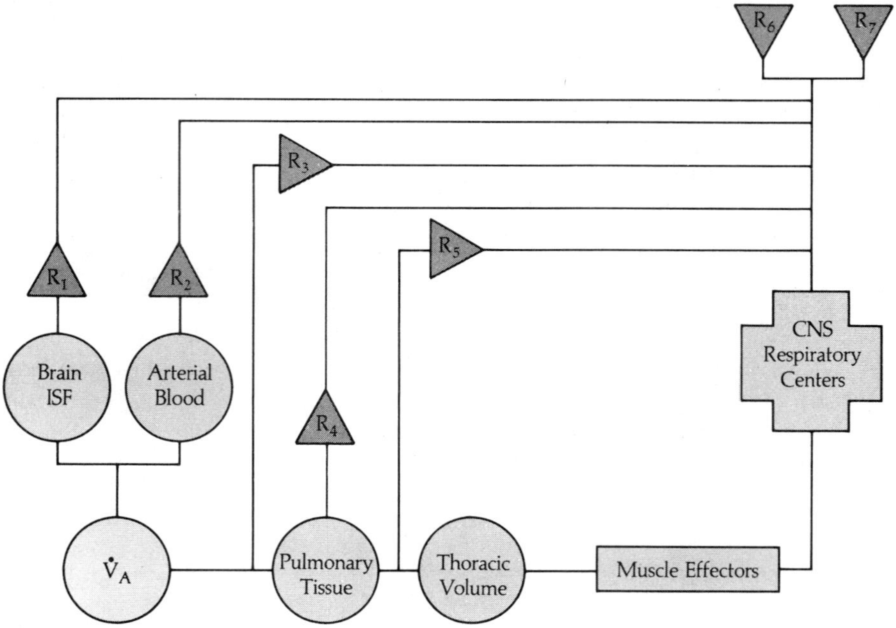

FIGURE 4-1 A model of the respiratory control system. *R* refers to receptors. Information from each set of receptors converges in the CNS to feed input to the respiratory centers. The respiratory centers in turn control the muscles of respiration (effectors). Changes in thoracic volume, pulmonary tissue volume, and composition of arterial blood and CSF are altered by changes in strength and frequency of muscle contractions. Information about changes in progress as well as those already completed is detected by the receptors and then relayed back to the CNS. R_1 is central chemoreceptors; R_2, peripheral chemoreceptors; R_3, pulmonary stretch receptors; R_4, other pulmonary receptors; R_5, stretch receptors in respiratory muscles; R_6, baroreceptors in carotid sinus and aortic arch; R_7, other muscle and joint receptors.

but they are depressed by hypoxia.* When the P_{CO_2} of medullary ISF increases, these receptors increase the rate at which they stimulate the respiratory centers in the medulla and ventilation ($\dot{V}_E$) increases. Central chemoreceptors are responsible for about 80% of the steady-state response to inspired CO_2. It is not known for certain whether the CO_2 acts directly on the chemoreceptor or the stimulus comes from H^+ formed as a result of hydration of CO_2 to carbonic acid. The current view is that CO_2 acts indirectly by means of H^+.

*The composition of the ISF of brain tissue is influenced by the composition of *both* CSF and blood. Frequently, it is stated that the central chemoreceptors respond to changes in P_{CO_2} of cerebral fluid (CSF). This is true in that when P_{CO_2} of CSF is altered the P_{CO_2} of brain and specifically medullary ISF is also changed in the same direction. However, it must be remembered that it is the medullary ISF that is the environment of central chemoreceptors. These chemoreceptors respond to changes in the composition of medullary ISF, not changes in CSF per se.

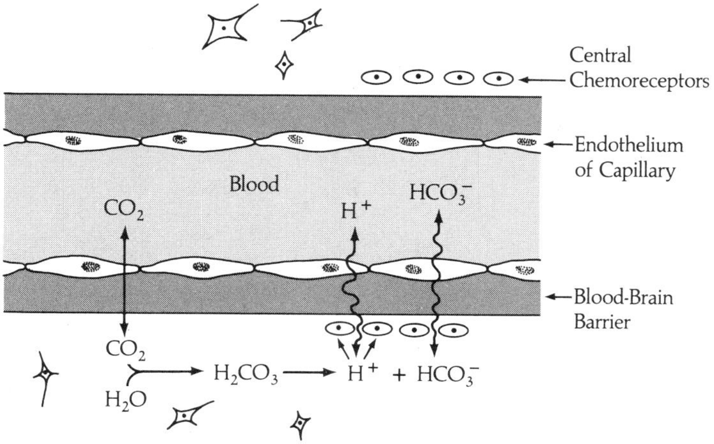

FIGURE 4-2 The central chemoreceptors and blood-brain barrier. The central chemoreceptors are located in the medullary brain tissue and are surrounded by ISF of brain tissue, which is in equilibrium with CSF and plasma. The structure of the chemoreceptors has not been described, hence, they are shown here diagramatically as ovoid-shaped cells. They may be complex neurons. The exact nature and structure of the blood-brain barrier has not been described either. The straight vertical arrow indicates rapid equilibration for CO_2 and the wavy arrows indicate slow equilibration.

Whatever the exact stimulus, the central chemoreceptors are very sensitive to P_{CO_2} changes in their environment.

Carbon dioxide in cerebral capillary blood equilibrates very rapidly with brain ISF. Therefore, changes in P_{CO_2} of medullary ISF will parallel those of arterial blood. However, changes in $[HCO_3^-]$ and pH of brain ISF from the addition of bases and fixed acids to blood are not as rapidly achieved as those for P_{CO_2}. The reason for the slower response to changes in $[HCO_3^-]$ and $[H^+]$ is the presence of the *blood-brain barrier* (Figure 4-2). The exact anatomic nature of the barrier is not known, but it does restrict movement of many types of molecules including H^+, HCO_3^-, antibiotics, and chemotherapeutic agents. The barrier has much less effect on CO_2 equilibration and, therefore, central chemoreceptors respond much more quickly to changes in blood P_{CO_2} than changes in blood $[HCO_3^-]$ or $[H^+]$ per se.

When the H^+ concentration of medullary ISF increases at constant or even reduced P_{CO_2}, the central chemoreceptors will stimulate an increase in ventilation rate. If fixed acids are added to the blood, the H^+ ions dissociated from those acids cross the blood-brain barrier slowly; hence, it may be several hours before the central chemoreceptors are stimulated significantly by the addition of the acid. However, ventilation rate increases very soon (minutes) after addition of fixed acid.

The stimulus that initially increases ventilation rate from addition of fixed acids is mediated by means of peripheral chemoreceptors (discussed below).

The increased ventilation rate will decrease arterial P_{CO_2} (Equation 4-9). When arterial P_{CO_2} is reduced, the P_{CO_2} of brain ISF is also reduced. Because of the decreased P_{CO_2} in brain ISF, $[H^+]$ decreases, hence the pH of brain ISF increases due to the reaction shown in Equation 4-10:

$$\underset{\substack{\downarrow \\ \text{removed from brain ISF}}}{CO_2} + H_2O \underset{2}{\overset{1}{\rightleftarrows}} H_2CO_3 \underset{4}{\overset{3}{\rightleftarrows}} H^+ + HCO_3^- \qquad (4\text{-}10)$$

The loss of CO_2 from the ISF of the medulla causes the rate of hydration of CO_2 (Reaction 1) to be reduced in relation to the rate of dehydration (Reaction 2). As a result, dissociation (Reaction 3) of carbonic acid is slowed compared to association (Reaction 4). Therefore, H^+ concentration in medullary ISF decreases until equilibrium is restored. Until H^+ from the fixed acid that was added to blood equilibrates with ISF across the blood-brain barrier, the pH of brain ISF will be increased while that of blood is decreased. During the period of time that the pH of medullary ISF is increased (P_{CO_2} decreased), the central chemoreceptors will not be stimulated as much as normal. Therefore, ventilation rate, although increased because of stimulation of peripheral chemoreceptors, will not be as great as would have occurred had the P_{CO_2} of brain ISF remained constant (Figure 4-3). Once H^+ equilibrates across the blood-brain barrier, the central chemoreceptors will be stimulated and $\dot{V}_A$ will be increased compared to the initial or early response of added fixed acid.

If after a day or two of equilibration the fixed acids were neutralized by intravenous infusion of $NaHCO_3$, blood pH would increase and stimulation of peripheral chemoreceptors would be reduced. However, the HCO_3^- infused would not cross the blood-brain barrier rapidly. Any reduction in alveolar ventilation from decreased stimulation of peripheral chemoreceptors would

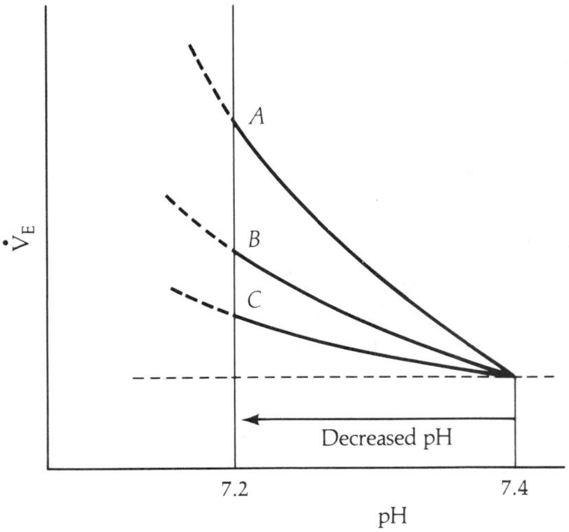

FIGURE 4-3
Respiratory response to added acid. Curve *A* is the response to hypercapnia. This response to increased plasma CO_2 is the result of stimulation of both central and peripheral chemoreceptors. Curve *B* is the response to addition of fixed acid at a constant P_{CO_2}. The response in *B* is due to stimulation of peripheral chemoreceptors alone. Curve *C* shows hyperventilation and, therefore, decreased arterial P_{CO_2} caused by fixed acids added to blood. The increased ventilation rate seen in *C* is due to stimulation of the peripheral chemoreceptors alone. The increment in $\dot{V}_E$ in *C* is less than that for *B* because the decreased P_{CO_2} reduces stimulation of the central chemoreceptors. Modified from *Respiratory Physiology*, 4th Ed., by N. B. Slonim and L. H. Hamilton. Copyright © 1981 by The C. V. Mosby Company. Reprinted by permission.

tend to increase the P_{CO_2} of both arterial blood and medullary ISF. If the P_{CO_2} of brain ISF rises without a concomitant rise in $[HCO_3^-]$, the central chemoreceptors would be stimulated and ventilation would return to the formerly high value. Therefore, when an acidosis from accumulation of fixed acids is corrected by infusion of HCO_3^-, the correction of the pH of medullary ISF lags behind that of blood because of the slow equilibration of HCO_3^-. As a result, alveolar ventilation will be maintained at a higher than normal rate, that is, hyperventilation, until the HCO_3^- in blood can equilibrate with brain ISF. The consequences of this lag in equilibration of HCO_3^- between blood and ISF of the medulla are explored further in Chapter 6.

Peripheral Chemoreceptors The peripheral chemoreceptors respond to changes in pH, P_{CO_2}, and P_{O_2} of arterial blood. The peripheral chemoreceptors (carotid and aortic bodies) are located at the dendritic terminals of cranial nerves IX and X. It should be noted that these chemoreceptors do not monitor blood-gas composition directly. Each receptor is surrounded by ISF, and it is the change in the composition of the ISF of the receptor that is detected. The chemoreceptors receive a huge blood flow for their size. It has been estimated that the carotid bodies receive 20 ml/minute of blood flow per gram of tissue.* This high flow causes the ISF of chemoreceptors to have essentially the same gas composition as that of arterial plasma. The flow is so high that oxygen requirements of receptors are met by physically dissolved O_2 in blood. Therefore, because of the high blood flow, changes in P_{O_2}, P_{CO_2}, and pH of arterial plasma will be quickly followed by changes in P_{O_2}, P_{CO_2}, and pH of the ISF of these receptors.

An increase in arterial P_{CO_2} will stimulate the peripheral chemoreceptors, which then in turn cause an increase in ventilation rate. The peripheral chemoreceptors contribute about 20% of the response to steady-state increased plasma P_{CO_2} (hypercapnia) at normal values of arterial P_{O_2}. However, the sensitivity of the peripheral chemoreceptors to changes in P_{CO_2} increases during hypoxemia (Figure 4-4). The sensitivity of the peripheral chemoreceptors to CO_2 decreases slightly during sleep and is sharply decreased with general anesthesia, chronic obstructive pulmonary disease (COPD), and narcotics (Figure 4-5).

Hypoxemia stimulates the peripheral chemoreceptors (Figure 4-6) but not central chemoreceptors. The central nervous system (CNS) in general and the respiratory centers in particular are depressed by hypoxemia. The more severe the hypoxemia, the more the CNS is depressed. At low values of arterial P_{O_2} the peripheral chemoreceptors may provide the major drive for respiration.

*The carotid bodies do not weigh even 0.1 g. The values for blood flow are given per gram of tissue to normalize the values for comparison to other tissues. For example, the kidneys receive the highest blood flow for a single organ at 4 to 6 ml/minute · g tissue. Because the chemoreceptors are not organs per se, they receive the highest flow for any tissue.

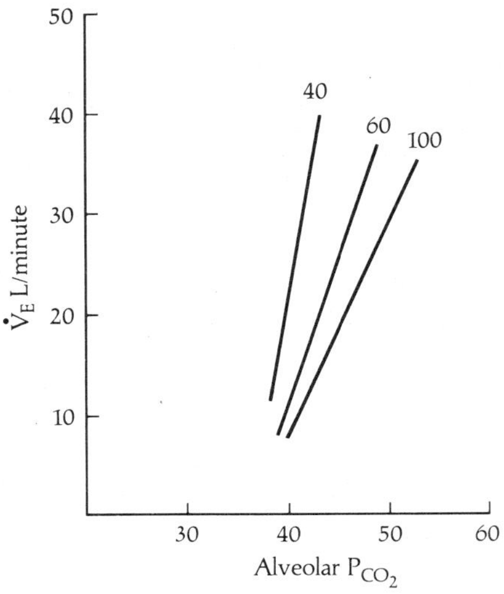

FIGURE 4-4
CO_2 response curves at three values of arterial P_{O_2}. Increasing alveolar P_{CO_2} by breathing CO_2 gas mixtures causes an increase in $\dot{V}_E$. The increased alveolar P_{CO_2} causes arterial P_{CO_2} to increase and results in stimulation of the chemoreceptors. The numbers at the top of each curve are values for arterial PO_2. Note the increased sensitivity to CO_2 as PO_2 decreases. The increased sensitivity is mediated by the peripheral chemoreceptors.

This is especially important in patients with COPD because the sensitivity of their chemoreceptors to CO_2 is decreased.

The response to hypoxia is potentiated by hypercapnia. At a normal P_{CO_2} in arterial blood the peripheral chemoreceptors do not generate much of an increase in ventilation until the P_{O_2} of arterial blood is decreased to approximately 60 torr. However, if arterial P_{CO_2} is increased above normal, then the peripheral chemoreceptors are more sensitive to hypoxemia (Figure 4-6). At higher values of arterial P_{CO_2}, a decrement in P_{O_2} of only 10 to 20 torr will cause an increase in ventilation rate.

Hydrogen ions directly stimulate both peripheral and central chemoreceptors. In Figure 4-3 the ventilatory response to added H^+ is shown as a function

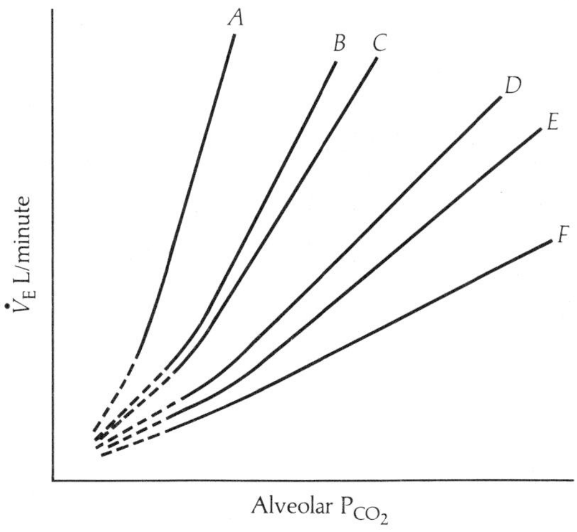

FIGURE 4-5
CO_2 response curves under different conditions: (*A*) metabolic acidosis, (*B*) normal awake response, (*C*) sleep, (*D*) narcotics, (*E*) chronic obstructive pulmonary disease, and (*F*) surgical anesthesia. Adapted from *Pulmonary Physiology*, by M. G. Levitzky. Copyright © 1982 by McGraw-Hill Book Company. Reprinted by permission.

FIGURE 4-6
Ventilatory responses to changes in P_{O_2} at three different values of arterial P_{CO_2} — 40, 46, and 50 torr. The sensitivity of peripheral chemoreceptors to hypoxia increases as arterial P_{CO_2} increases.

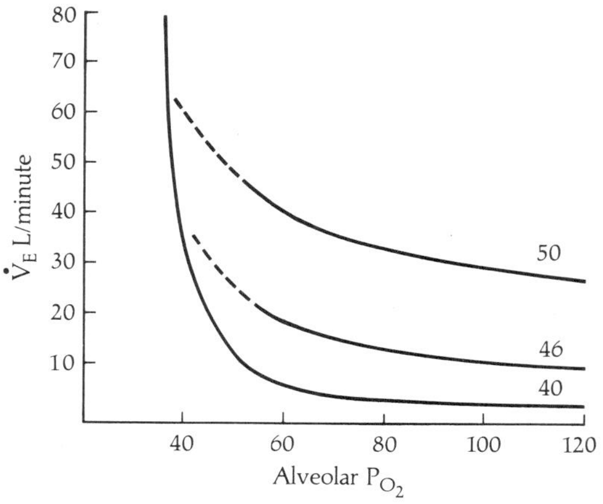

of pH. The H^+ concentration is increased in curve *A* by adding CO_2, thus, P_{CO_2} increases at all points along curve *A* as pH is reduced from 7.4 to 7.2. Curve *C* is typical of the early response of the respiratory system to addition of nonvolatile acids. In acidosis of nonrespiratory origin, the initial increase in ventilation rate must come from the peripheral chemoreceptors because the central chemoreceptors will not be stimulated until sufficient time has passed for significant transport of H^+ across the blood-brain barrier. After equilibration of H^+ between plasma and brain ISF, ventilation rate will be increased over the initial increment and arterial P_{CO_2} will be decreased more than with peripheral chemoreceptor stimulation alone.

In summary, regulation of arterial P_{CO_2} requires regulation of alveolar ventilation rate. The regulation of ventilation is mediated through the CNS. Sensory input from chemoreceptors sensitive to changes in P_{CO_2}, $[H^+]$, and P_{O_2} provides information about arterial blood-gas composition as well as pH and P_{CO_2} of the ISF of brain tissue. Disease processes that reduce alveolar ventilation, such as COPD, severe asthma, severe pneumonia, and pulmonary edema will increase arterial P_{CO_2}. Several events can cause hyperventilation, such as metabolic acidosis, fever, emotional disorders, and hypoxia of high altitude. Each of these events can cause hypocapnia. It is also important in our modern society to be aware that head injuries and spinal cord injuries can cause hypoventilation and even apnea, which lead to increased arterial P_{CO_2} as well as reduced arterial P_{O_2}.

MATCHING VENTILATION WITH PERFUSION

In healthy adults the alveolar space is composed of approximately 300 million alveoli. Up to this point in the discussion it has been implicitly assumed that ventilation of these alveoli is matched with the blood flow to these alveoli. This is more or less the case in health. There are local reflexes that adjust blood flow to match ventilation. However, ventilation and blood flow can be mismatched in respiratory diseases such as COPD and adult respiratory distress syndrome (ARDS). With mismatching, proportionally more blood flows to alveoli that are less well-ventilated and vice versa. Mismatching of ventilation with perfusion can cause CO_2 retention and, therefore, hypercapnia. If the degree of mismatching is mild, hypercapnia may not occur because the central chemoreceptors increase overall ventilation to keep arterial P_{CO_2} constant. If the degree of mismatching is more pronounced, as can occur in COPD and ARDS, then hypercapnia can develop.

The effect of mismatching ventilation and blood flow on arterial P_{O_2} is much more pronounced than it is on P_{CO_2}. When arterial P_{CO_2} starts to increase from mismatching, increased alveolar ventilation from stimulation of chemoreceptors will decrease the CO_2 content of blood flowing to well-ventilated alveoli until mixed arterial blood has a P_{CO_2} of 40 torr. However, with O_2, increasing ventilation to already well-ventilated alveoli will not improve O_2 saturation more than 2 to 3% because blood flowing to well-ventilated regions is already nearly saturated with O_2 (97–98%), as shown in Figure 3-4. Consequently, when blood from less well-ventilated regions mixes in the left ventricle with blood from well-ventilated regions, the P_{O_2} cannot "average out" to normal values. Hence, P_{O_2} of arterial blood always decreases with mismatching of ventilation and blood flow. The hypoxemia that results can cause acid-base disturbances (see Chapter 9).

RENAL REGULATION OF PLASMA HCO_3^- CONCENTRATION

The kidneys regulate bicarbonate concentration through three processes in normal adults. These processes share a common mechanism involving secretion of H^+ into tubular fluid across the luminal membrane (Figure 4-7). The H^+ is

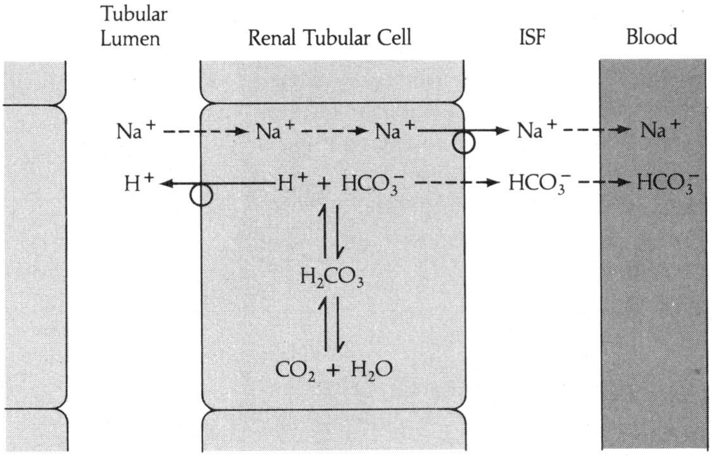

FIGURE 4-7
The basic mechanism of generating HCO_3^- in renal tubular cells. Sodium is exchanged for hydrogen ions at the luminal border whereas HCO_3^- accompanies reabsorbed sodium ions into the blood at the peritubular border. It is through this basic mechanism that the kidneys maintain normal HCO_3^- concentration and also compensate for changes in acid-base balance. The solid arrows with the circle represent active transport. *Source:* "Basic mechanisms involved in acid-base homeostasis," by J. L. Keyes, *Heart and Lung*, 1976, 5:239–246. Reprinted by permission of The C. V. Mosby Company.

generated in the cells of the proximal and distal convoluted tubules from carbonic acid. To maintain electroneutrality, Na^+ is taken into the cell in exchange for the secreted H^+. The HCO_3^- that is generated along with the H^+ from dissociation of carbonic acid in the cell then accompanies the sodium across the peritubular membrane to be reabsorbed into the blood.

The three processes regulating HCO_3^- concentration are (1) "reabsorption" of filtered HCO_3^-, (2) formation of titratable acidity, and (3) formation of ammonium ions (NH_4^+). All three processes require secretion of H^+ and each process is discussed in more detail in the following sections.

Reabsorption of Filtered HCO_3^- The kidneys filter nearly 180 L of plasma water per day. In that filtered volume there are about 4300 mEq of $NaHCO_3$, that is, 180 L/day × 24 mEq/L = 4320 mEq/day. A person on a normal diet recovers or replaces all of the HCO_3^- that is filtered. The process is shown in Figure 4-8. Note that the HCO_3^- that accompanies the sodium ion reabsorbed into peritubular blood is not the same HCO_3^- filtered with that sodium cation.

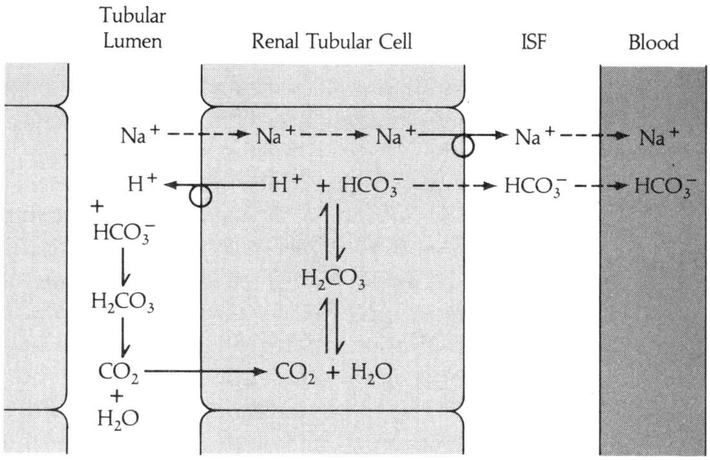

FIGURE 4-8
Processes for "reabsorption" of filtered HCO_3^-.

Bicarbonate reabsorption is more a process of replacement of filtered HCO_3^-, ion for ion, than reabsorption per se. However, the overall effect is the same as if the filtered HCO_3^- had been reabsorbed with sodium. This process occurs throughout the nephron except for the descending limb of Henle's loop. Most occurs in the proximal tubule (80–90%) whereas the remainder is reabsorbed in the ascending limb of Henle's loop and the distal tubule. Therefore, normally all filtered HCO_3^- is reabsorbed.

Formation of Titratable Acidity There are conjugate bases other than HCO_3^- for H^+ to associate with in tubular fluid. This is especially important once all filtered HCO_3^- has been reabsorbed. Phosphate becomes concentrated

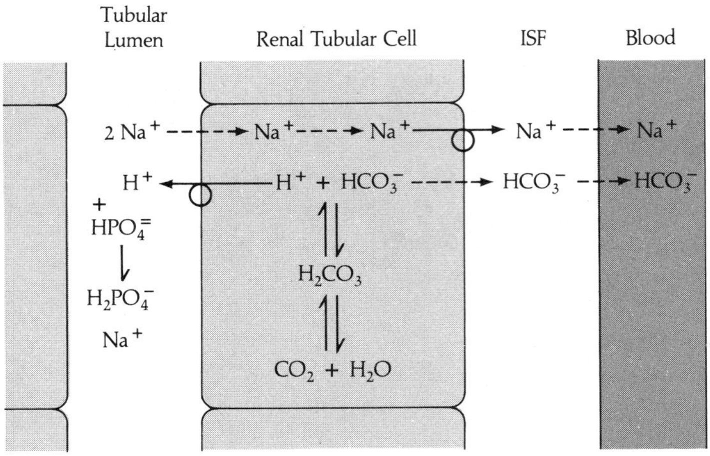

FIGURE 4-9
Processes involved in the formation of titratable acidity.

in tubular fluid because only about 75% of the filtered phosphate is reabsorbed. The monohydrogen form of phosphate shown in Figure 4-9 is accompanied in tubular fluid by Na^+. The kidney must recover some of this Na^+ to regulate the volume of extracellular fluid. Sodium is exchanged for H^+ across the luminal membrane and is again accompanied by HCO_3^- across the peritubular membrane. The H^+ that was secreted combines with the monohydrogen phosphate to form dihydrogen phosphate. The dihydrogen phosphate is excreted in the urine.

The HCO_3^- accompanying Na^+ into peritubular blood is new HCO_3^- that is added to extracellular fluid. Thus, the amount of HCO_3^- leaving the kidney in the renal vein is greater than that entering via the renal artery. *The kidneys are, thus, HCO_3^- generators.* This added HCO_3^- replaces bicarbonate lost in buffering fixed acids from metabolism (Chapter 5). In other words, excreted phosphate is replaced in the blood with HCO_3^-.

The term *titratable acid* comes from the laboratory method used to determine the amount of HCO_3^- generated by titrating phosphate in tubular fluid. A sample of urine is titrated with NaOH to return the pH to that of the patient's arterial blood (7.4 in normal people). The base titrates the phosphate as shown in Equation 4-11:

$$OH^- + H_2PO_4^- \rightleftarrows HPO_4^= + H_2O \qquad (4\text{-}11)$$

The amount of OH^- (in milliequivalents) required to titrate the urine back to arterial pH is equal to the amount of H^+ ion (in milliequivalents) secreted and, therefore, to the amount of new HCO_3^- generated from titrating phosphate in tubular fluid. Because the acid $H_2PO_4^-$ is titrated, the term titratable acidity is used. It should be noted that other buffers are present and also participate in the reactions shown in Figure 4-9; however, phosphate is quantitatively the most important. Normally, approximately 40 mEq of phosphate are excreted per day, and about half is in the dihydrogen form. Therefore, about 20 mEq of new HCO_3^- are added to blood each day from formation of titratable acidity.

Formation of Ammonium Ions Renal tubular cells synthesize ammonia (NH_3) from glutamine. Ammonia is a relatively strong conjugate base in tubular fluid and quickly associates with H^+ to form NH_4^+ (Figure 4-10). Again, Na^+ is exchanged for H^+ in this process at the luminal border of the cell and the Na^+ accompanies HCO_3^- into capillary blood across the peritubular border. As was the case for phosphate, the anion accompanying Na^+ in the tubular fluid is replaced with HCO_3^- when the Na^+ is reabsorbed. Once the secreted H^+ combines with ammonia, it remains trapped in the tubular fluid and is not reabsorbed. Thus, formation of NH_4^+ ion permits addition of more new HCO_3^- to blood. Again this new HCO_3^- replaces that lost to buffering fixed acids from normal metabolism, as is explained in Chapter 5. Normally, the kidneys generate from 30 to 40 mEq of HCO_3^- per day from NH_4^+ ion formation.

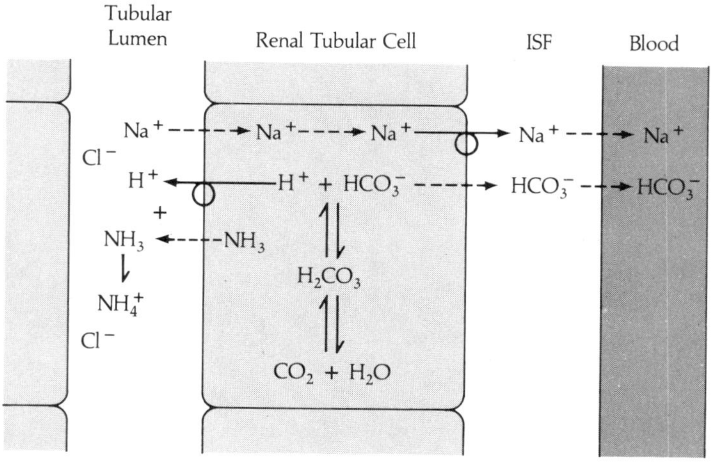

FIGURE 4-10
Processes involved in the formation of ammonium (NH_4^+) ion.

The kidneys are HCO_3^- generators. They replace all of the filtered HCO_3^- normally (4300 + mEq/day) and generate new HCO_3^- (50–60 mEq/day) to replace that used in buffering fixed acids. It is true that the kidneys excrete acid in the urine ($H_2PO_4^-$ and NH_4^+). However, it is important to recognize that the hydrogen ions associated with the $HPO_4^=$ and NH_3 bases were never in the extracellular fluid per se; rather they were generated in the kidney tubular cells and secreted into the tubular fluid where they were combined with bases. Hence, these hydrogen ions were not removed from ECF. The kidneys add HCO_3^- to ECF as a part of maintaining fluid and electrolyte balance.

RENAL HCO_3^- SYNTHESIS AND REGULATION OF ECF VOLUME

The regulation of HCO_3^- concentration is closely intertwined with sodium reabsorption, which in turn is a principal process for regulation of ECF volume. Two interrelated factors that are part of ECF volume regulation can affect HCO_3^- generation by the kidneys: (1) the renin-angiotensin system and (2) Na^+ and Cl^- delivery to the distal nephron.

The Renin-Angiotensin System The renin-angiotensin control system is described in Chapter 2, and a diagram of the feedback control system is shown in Figure 2-7. A major action of the system is to regulate the synthesis and release of aldosterone, which in turn regulates reabsorption of sodium by the kidney.

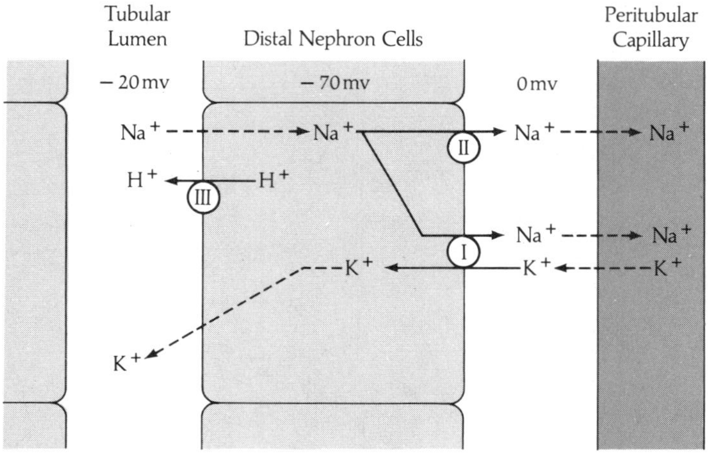

FIGURE 4-11 The relationship between Na^+ reabsorption and H^+ and K^+ secretion. The circles with solid arrows represent active transport and the dashed arrows represent simple diffusion. Aldosterone stimulates active transport pump I, increasing Na^+ reabsorption and K^+ uptake by the cell. Aldosterone may also stimulate pump III. Either K^+ or H^+ may exchange for Na^+ at the luminal border of the cell. Pump II is a Na^+ transport pump that is not stimulated by aldosterone and does not exchange Na^+ for K^+. Pump III is the H^+ secretory pump. The potentials shown are given in reference to the zero reference potential in the ISF between the peritubular capillary and distal nephron cells.

Specifically, aldosterone stimulates the Na^+-K^+ exchange pump in cells of the distal nephron. The net result is an increase in the rate of reabsorption of Na^+ and increased K^+ secretion (Figure 4-11). Aldosterone also increases H^+ secretion indirectly by increasing the rate of Na^+ reabsorption and directly by stimulating H^+ secretion itself. The kidney always generates a HCO_3^- ion when it secretes H^+. This HCO_3^- then is reabsorbed along with Na^+, as shown in Figures 4-7 through 4-10. However, it should be noted that aldosterone itself will generally not increase the HCO_3^- concentration in plasma more than 3 to 4 mEq/L in humans, even when the aldosterone concentration is much greater than normal. To increase $[HCO_3^-]$ significantly with increased aldosterone secretion, other factors must also be present such as a chloride deficit. This is explained further in the next section.

Sodium and Chloride Delivery to Distal Nephron When the amount of sodium delivered from the proximal tubule and Henle's loop to the distal nephron increases, the rate of reabsorption of Na^+ increases. A model of the steps involved is shown in Figure 4-12. As more sodium enters the cell, the lumen tends to become progressively more electronegative due to transport of cation out of tubular fluid. The increased electronegativity has several effects. First, it increases the rate of K^+ diffusion out of the cell because of the increased negative charge in the lumen. Second, the negative charge also accelerates H^+ secretion. When the Cl^- concentration of tubular fluid is lower than normal, as frequently occurs with diuretic therapy or with prolonged loss of gastric secretion, then more H^+ and K^+ will be exchanged for Na^+. When H^+ secretion increases, more HCO_3^- is generated and $[HCO_3^-]$ in plasma increases. Thus, K^+, Cl^-, Na^+, H^+, and HCO_3^- are all intertwined in regulation of ECF volume. It is to be expected that disease processes that alter fluid balance will also frequently alter acid-base regulation. Clinical examples of patients with volume and Cl^- deficits are given in Chapter 10.

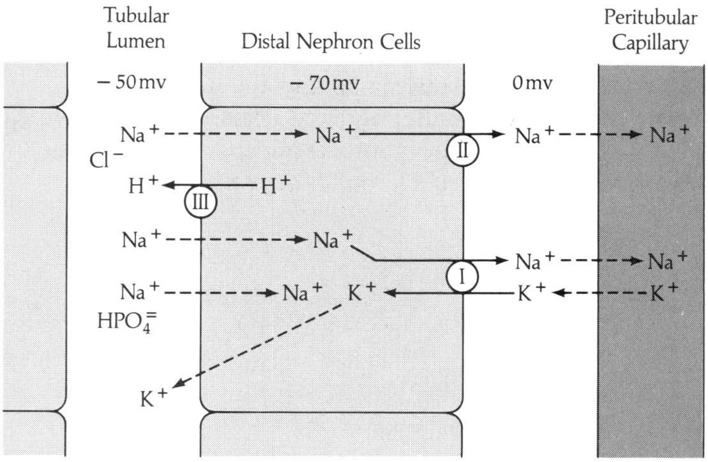

FIGURE 4-12
The effect of increased Na^+ load delivered to the distal nephron. The increased Na^+ load causes an increased reabsorption of Na^+ from the lumen into the cells. As a result the lumen becomes more electronegative (see Figure 4-11). The increased electronegativity may increase the rate of K^+ and H^+ secretion. Many kinds of diuretics may increase K^+ secretion in this way. The solid arrows with circles indicate active transport. The numbers in the pumps are described in Figure 4-11.

SUMMARY

Carbon dioxide content and, therefore, P_{CO_2} of blood are regulated by ventilation. Ventilation is precisely controlled by a complex feedback control system involving the central nervous system, peripheral chemoreceptors, and mechanoreceptors. Chemical factors stimulating ventilation include changes in pH

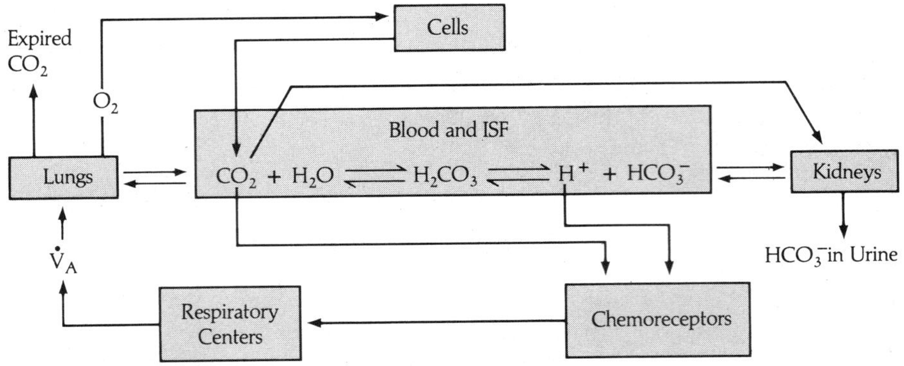

FIGURE 4-13 Interaction of the respiratory system and kidneys in regulating P_{CO_2} and $[HCO_3^-]$. CO_2 can be added to blood when needed by decreasing $\dot{V}_A$. Bicarbonate can be excreted when present in the blood in excess amounts. Ordinarily, HCO_3^- is generated by the kidneys and added to the blood. CO_2 can stimulate increased HCO_3^- generation by the kidneys.

and P_{CO_2} of blood and medullary ISF. In the normal individual, ventilation rate changes quickly to maintain arterial P_{CO_2} within the narrow normal range (38–42 torr). The respiratory system excretes about 300 L of CO_2 per day (15000 mEq of acid). Maintenance of P_{CO_2} also requires that blood flow and ventilation be properly matched.

Plasma bicarbonate concentration is regulated by the kidneys as a part of Na^+ as well as H^+ balance. In healthy individuals, the kidneys generate between 50 and 100 mEq of HCO_3^- per day. This HCO_3^- replaces that lost in buffering fixed acids produced from catabolism. Renal and respiratory regulatory processes are coupled through the blood and the $CO_2 - H_2CO_3 - HCO_3^-$ buffer system (Figure 4-13).

Buffering of Volatile and Nonvolatile Acids and Bases

CHAPTER 5

Buffering is the first line of defense for maintaining acid-base status. When acid-base disturbances are generated as a result of disease or trauma, buffering immediately minimizes the magnitude of change in pH of body fluids that would otherwise occur. For all practical purposes, buffering is instantaneous and occurs as the acid-base disturbance develops. A thorough knowledge of buffering processes is essential for understanding both acid-base physiology and pathophysiology. In addition, assessment of acid-base disturbances is based in part on changes in blood-gas composition resulting from buffering.

BUFFER PAIRS IN BODY FLUID COMPARTMENTS

Each body fluid compartment has its own set of buffer pairs (or buffer systems). In this text the major buffer systems are named as shown in Table 5-1. Only those buffers present in sufficient concentrations to be effective are listed.

Blood Buffers

Blood is made up of two compartments, plasma and red cells. There are three major buffer systems in blood (Table 5-1). The plasma proteins buffer both carbonic and fixed acids. Hemoglobin, found only in the red blood cell, buffers

Table 5-1 Major Buffer Systems in Body Fluid Compartments

Compartment	Buffer system*	Reaction
Blood (plasma)	Plasma proteins	$HPr \rightleftharpoons H^+ + Pr^-$
(Red cell)	Hemoglobin	$HHb \rightleftharpoons H^+ + Hb^-$
(Red cell and plasma)	Bicarbonate	$CO_2 + H_2O \rightleftharpoons H_2CO_3 \rightleftharpoons H^+ + HCO_3^-$
ISF	Bicarbonate	
ICF	Protein	$HPr \rightleftharpoons H^+ + Pr^-$
	Phosphate	$H_2PO_4^- \rightleftharpoons H^+ + HPO_4^=$

The conjugate acids of plasma proteins, hemoglobin, cell proteins, and phosphate buffer systems are anions in body fluids.

volatile acid during transport of CO_2 in normal respiration as well as during abnormal changes in CO_2 content of blood. Hemoglobin also buffers H^+ shifting between red cells and plasma during acid-base disturbances. The bicarbonate buffer system buffers fixed acids produced from metabolism.

Buffers in ISF

The only buffer system of significance found in ISF is the bicarbonate buffer system. Although phosphate is found in both plasma and ISF, it is present in such low concentrations that it is not an effective buffer in these compartments. There are essentially no protein buffers in ISF.

Buffers in ICF

Both proteins and phosphate are major intracellular buffers. Phosphate is found in high concentrations in the ICF and thus is an important buffer. The bicarbonate buffer system, although present in the ICF, is not as important as phosphate and protein for intracellular buffering.

REACTIONS IN BUFFERED SOLUTIONS

Suppose HCl is added to distilled water to make up 1 L of HCl solution in a concentration of 100 mmol/L (0.1 mol/L). The chemical reaction is described in Equation 5-1:

$$HCl \rightarrow H^+ + Cl^- \quad (5\text{-}1)$$

The $[H^+]$ and pH may be determined as follows:

Step 1 HCl is a strong acid and is, therefore, essentially completely dissociated. The liter of solution will contain 100 mmol each of H^+ and Cl^-. Hence, $[H^+]$ is 100 mmol/L or 0.1 mol/L.

Step 2 pH can be calculated from Equation 5-2:

$$\text{pH} = -\log[\text{H}^+] = \log\frac{1}{[\text{H}^+]} \quad (5\text{-}2)$$

$$\text{pH} = \log\frac{1}{0.1} = \log 10 = 1.0 \quad (5\text{-}3)$$

Thus, when $[H^+]$ is known, the pH can be easily calculated. If a beaker contains a buffered solution, addition of 100 mmol of HCl will cause the pH to decrease; however, the presence of the buffer makes determination of pH more complicated. The pertinent reactions are shown in Equations 5-1 and 5-4:

$$\text{HB} \rightleftharpoons \text{H}^+ + \text{B}^- \quad (5\text{-}4)$$

The interaction of the two reactions can be seen more clearly if Equations 5-1 and 5-4 are written crossing at the common ion, H^+:

$$\begin{array}{c} \text{HCl} \\ \downarrow \\ \text{HB} \rightleftharpoons \text{H}^+ + \text{B}^- \\ + \\ \text{Cl}^- \end{array} \quad (5\text{-}5)$$

The association component of the horizontal reaction in Equation 5-5 acts to absorb *some* of the added H^+ from the HCl (buffering). The net result is that $[H^+]$ increases because of the added HCl but not as much as with distilled water. The actual pH may be determined by physical measurement using a pH electrode or calculated as described in the following section.

DERIVATION OF THE HENDERSON AND HENDERSON-HASSELBALCH EQUATIONS*

Equation 5-4 describes the equilibrium reaction of a weak acid in solution. If the velocity of dissociation is V_1 and velocity of association is V_2, then, as discussed under the section on equilibrium in Chapter 1,

$$V_1 = k_1\,[\text{HB}] \quad (5\text{-}6)$$

$$V_2 = k_2\,[\text{H}^+][\text{B}^-] \quad (5\text{-}7)$$

*The equations derived in this section are included for the more serious student of acid-base regulation. They are not very difficult to follow; however, some find that these derivations are more meaningful after they have a better understanding of buffering. It is left up to the reader to choose whether to study this section for more depth or to proceed directly to Equation 5-27. Once general understanding of buffering is achieved, the reader may wish to return to review this section in more detail.

At equilibrium: $$V_1 = V_2 \tag{5-8}$$

$$k_1[HB] = k_2\,[H^+][B^-] \tag{5-9}$$

rearranging $$\frac{k_1}{k_2}[HB] = [H^+][B^-] \tag{5-10}$$

and thus $$\frac{k_1}{k_2} = \frac{[H^+][B^-]}{[HB]} = Ka \tag{5-11}$$

where *Ka* is the dissociation constant. The larger the value of *Ka*, the more readily dissociation goes to completion. The Henderson equation is obtained by rearranging Equation 5-11:

$$[H^+] = Ka\frac{[HB]}{[B^-]} \tag{5-12}$$

When the concentrations of HB and B^- are known, *Ka* may be obtained from a table of dissociation constants and $[H^+]$ may be calculated. The Henderson equation is sometimes used when the value for $[H^+]$ is desired.

Usually the value of pH is used, and it may be obtained by modifying the Henderson equation. First, take the logarithm of both sides of Equation 5-12,

$$\log[H^+] = \log Ka\frac{[HB]}{[B^-]} \tag{5-13}$$

The log of the product of two numbers is equal to the sum of the logarithms of the two numbers, hence,

$$\log Ka\frac{[HB]}{[B^-]} = \log Ka + \log\frac{[HB]}{[B^-]} \tag{5-14}$$

Substituting Equation 5-14 into 5-13,

$$\log\,[H^+] = \log Ka + \log\,\frac{[HB]}{[B^-]} \tag{5-15}$$

multiplying by (-1)

$$-\log[H^+] = -\log Ka - \log\,\frac{[HB]}{[B^-]} \tag{5-16}$$

Since $-\log\,[H^+]$ is defined as pH (Equation 5-2),

$$pH = \log\frac{1}{Ka} + \log\frac{1}{[HB]/[B^-]} \tag{5-17}$$

The term, $\log \frac{1}{Ka}$, is called p*Ka*,* thus,

$$\text{pH} = \text{p}Ka + \log \frac{[\text{B}^-]}{[\text{HB}]} \quad (5\text{-}18)$$

Equation 5-18 is the Henderson-Hasselbalch equation. The values of p*Ka* for different buffers are available from tables in chemistry handbooks. When the values for $[B^-]$ and [HB] are known, pH may be calculated from Equation 5-18. It can also be seen from Equation 5-18 that if the ratio $[B^-]/[HB]$ increases, pH increases or, conversely, if the ratio of $[B^-]/[HB]$ decreases, pH decreases. Originally, the Henderson-Hasselbalch equation permitted calculation of pH by measuring the concentrations of conjugate acid-base pairs. After the glass pH electrode came into general use, pH could be measured directly. If pH is measured using the glass electrode, then measurement of the concentration of one of the buffer pairs permits calculation of the other member of the pair. This application of Equation 5-18 is explained further in the next section.

THE HENDERSON-HASSELBALCH EQUATION APPLIED TO THE HCO_3^- BUFFER SYSTEM

There are many buffer pairs in blood; however, there can be only one value for $[H^+]$ at equilibrium (isohydric principle). All buffer pairs are in equilibrium with the same H^+ pool, therefore:

$$[\text{H}^+] = Ka_1 \frac{[\text{HB}_1]}{[\text{B}_1]} = Ka_2 \frac{[\text{HB}_2]}{[\text{B}_2]} = Ka_3 \frac{[\text{HB}_3]}{[\text{B}_3]} = \cdots = Ka_n \frac{[\text{HB}_n]}{[\text{B}_n]} \quad (5\text{-}19)$$

Equation 5-19 shows that in a complex solution such as plasma, the $[H^+]$ (and, therefore, pH) is determined by the concentration ratio of buffer pairs present and their respective dissociation constants. Any change in acidity produced by addition of an acid or a base will cause a redistribution of H^+ among all buffer pairs until equilibrium is reestablished. From Equation 5-19 it can be seen that to determine pH of a solution with multiple buffers, we need to measure the concentrations of only a single buffer pair and substitute the values obtained into Equation 5-18.

To assess acid-base status in humans, the components of the HCO_3^- buffer system are measured. To determine pH of blood, the conjugate acid-base pair of the HCO_3^- buffer system is used in Equation 5-18. However, the form of the equation is somewhat different from that shown in Equation 5-18. The following derivation shows how the final form of the Henderson-Hasselbalch equa-

*The larger the value of p*Ka*, the weaker the acid is. The concept of p*Ka* is discussed more completely in Appendix 1.

tion for the HCO_3^- buffer system is obtained. The reaction between CO_2, HCO_3^-, and H^+ is

$$CO_2 + H_2O \rightleftharpoons H_2CO_3 \rightleftharpoons H^+ + HCO_3^- \tag{5-20}$$

From Equation 5-12 the Henderson equation for carbonic acid would be

$$[H^+] = Ka\frac{[H_2CO_3]}{[HCO_3^-]} \tag{5-21}$$

The concentration of H_2CO_3 is very difficult to measure; however, it is known that

$$[H_2CO_3] = \frac{[CO_2]}{340} \tag{5-22}$$

where $[CO_2]$ is the concentration of physically dissolved CO_2. Substituting Equation 5-22 into 5-21, we obtain,

$$[H^+] = \frac{Ka}{340} \cdot \frac{[CO_2]}{[HCO_3^-]} \tag{5-23}$$

Since Ka and 340 are both constants, the ratio of $Ka/340$ is also constant and is symbolized as Ka'. Thus,

$$[H^+] = Ka' \cdot \frac{[CO_2]}{[HCO_3^-]} \tag{5-24}$$

From Henry's law (Chapter 1)

$$CO_2 = S \cdot P_{CO_2} \tag{5-25}$$

where S is the solubility coefficient of CO_2 in plasma and is equal to 0.0301 mm CO_2/torr P_{CO_2} at 37°C. Therefore,

$$[H^+] = Ka'\frac{S \cdot P_{CO_2}}{[HCO_3^-]} \tag{5-26}$$

If steps of Equations 5-13 through 5-18 are repeated, Equation 5-26 becomes

$$pH = pKa' + \log\frac{[HCO_3^-]}{S \cdot P_{CO_2}} \tag{5-27}$$

where $pKa' = 6.1$ at 37°C.* Equation 5-27 is the Henderson-Hasselbalch equation commonly used for determining blood acid-base status.

---

*pKa' is affected by temperature and pH. For purposes of this text, it will be assumed that pKa' is 6.1. However, most modern laboratories correct for pH and temperature when blood-gas data are reported. The assumption of a constant value for pKa' will not alter the principles or conclusions presented in this text.

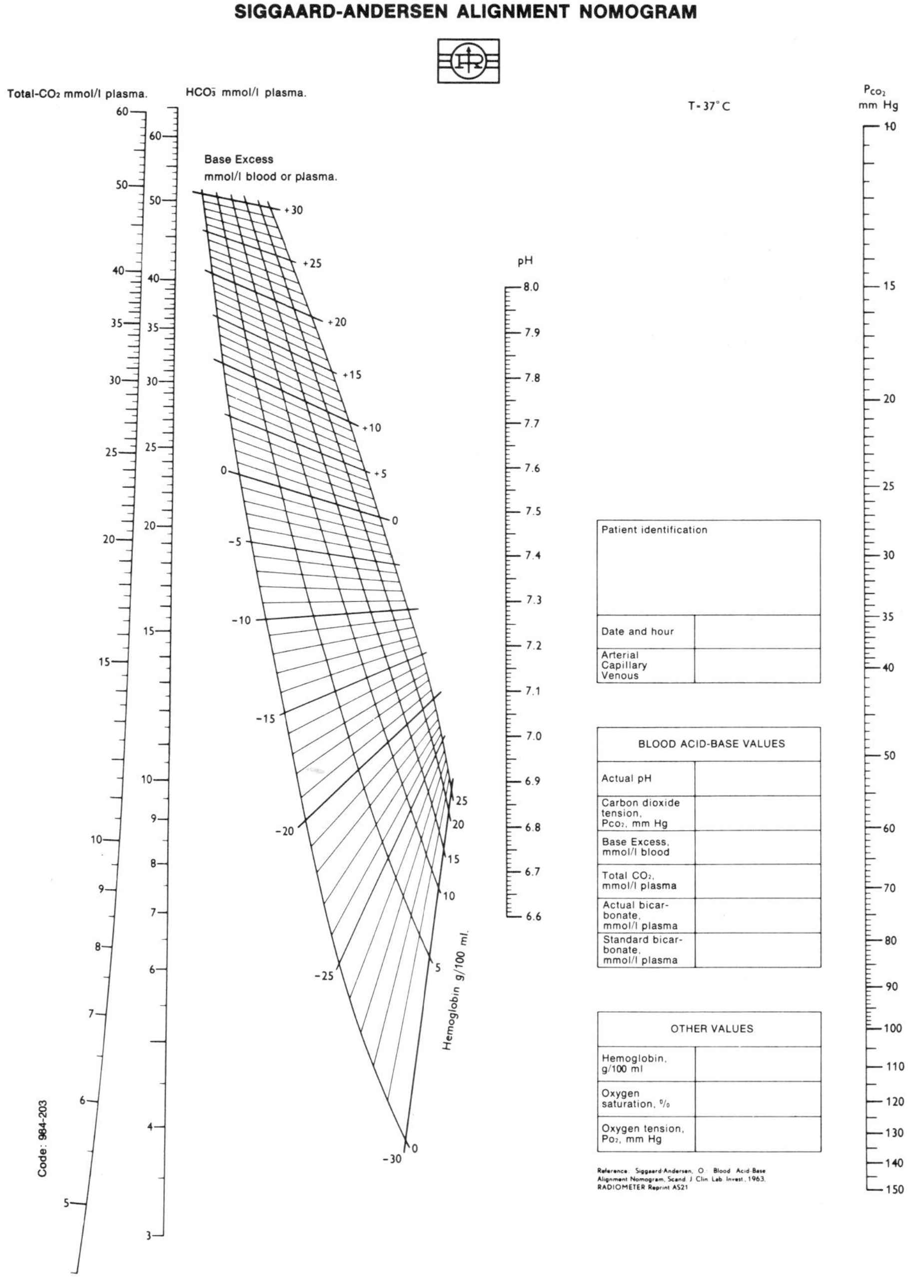

FIGURE 5-1 The Siggaard-Andersen Nomogram. Copyright © 1963 by Radiometer A / S, Copenhagen, Denmark. Reprinted by permission.

Generally, pH is not calculated; rather, pH and P_{CO_2} are measured and the $[HCO_3^-]$ is calculated by rearranging Equation 5-27:

$$[HCO_3^-] = [10^{pH - pKa}][S \cdot P_{CO_2}] \quad (5\text{-}28)$$

Equation 5-28 is awkward to use unless one is skilled using log tables or has a calculator available. Equation 5-28 can be solved using a nomogram such as the Siggaard-Andersen nomogram (Figure 5-1). To use the Siggaard-Andersen nomogram, values for pH and P_{CO_2} obtained from blood-gas analysis are marked as points on the appropriate scales of the nomogram. A straight line joining these two points is extended through the $[HCO_3^-]$ and total CO_2 scales. The $[HCO_3^-]$ for a given pH and P_{CO_2} is equal to the value obtained at the intersection of the straight line and $[HCO_3^-]$ scale. The total CO_2 is found at the intersection of the straight line and total CO_2 scale. Total CO_2 is equal to the sum of $[HCO_3^-]$ and physically dissolved CO_2.

Example: If the pH and P_{CO_2} are 7.32 and 54 torr, respectively, the $[HCO_3^-]$ is 27.3 mEq/L and total CO_2 is 29 mmol/L (Figure 5-1).

REACTIONS INVOLVING THE HCO_3^- BUFFER SYSTEM

Acid-base status is determined from the concentrations of the components of the CO_2-HCO_3^- buffer system. Both $[H^+]$ and $[HCO_3^-]$ change during an acid-base disturbance. However, these two parameters do not always change in the same direction. The pattern of changes that occurs depends on whether the initial event is a change in volatile acid, fixed acid, or base concentrations. The discussion that follows describes what happens to $[HCO_3^-]$ in plasma when $[H^+]$ increases and decreases.

Processes that Increase $[H^+]$

There are two general processes that increase the $[H^+]$ of body fluids: (a) addition of acid and (b) loss of HCO_3^-.

Addition of Acid It can be seen from Equation 5-29 that addition of either volatile or fixed acid to the ECF will cause an increase in the $[H^+]$ of blood.

$$CO_2 + H_2O \rightleftharpoons H_2CO_3 \underset{V_{a_1}}{\overset{V_{d_1}}{\rightleftharpoons}} \begin{matrix} HB \\ {\scriptstyle V_{a_2}} \upharpoonleft\!\downharpoonright {\scriptstyle V_{d_2}} \\ H^+ \\ + \\ B^- \end{matrix} + HCO_3^- \quad (5\text{-}29)$$

In this equation, V_a is the velocity of association and V_d the velocity of dissociation. HB represents a weak fixed acid and B^- its conjugate base. When CO_2 (volatile acid) is added to ECF, either through hypoventilation or increased $F_{I_{CO_2}}$, V_{d_1} increases and both $[H^+]$ and $[HCO_3^-]$ will increase. The increase in $[H^+]$ will also cause V_{a_2} to increase and $[B^-]$ to decrease. When equilibrium is reestablished after addition of CO_2 to the ECF, both $[H^+]$ and $[HCO_3^-]$ will be increased.

If the concentration of fixed acid, [HB], is increased, as occurs in ketoacidosis, then V_{d_2} will be temporarily increased, producing an increased concentration of both H^+ and B^-. Because $[H^+]$ increases, V_{a_1} increases, thereby decreasing $[HCO_3^-]$. Therefore, when equilibrium is achieved after addition of fixed acids to ECF, the $[H^+]$ will be increased (i.e., pH will be decreased) and the $[HCO_3^-]$ will be reduced.

Loss of HCO_3^- Loss of HCO_3^- as occurs in diarrhea will cause a temporary reduction in V_{a_1}. Therefore, until equilibrium is reestablished, $V_{d_1} > V_{a_1}$. The net result is an increase in $[H^+]$ and a decrease in $[HCO_3^-]$. It should also be noted that other buffers will also be affected by loss of HCO_3^-. The increased $[H^+]$ will cause V_{a_2} to increase, thereby causing a reduction in $[B^-]$ once equilibrium is restored.

Processes that Decrease $[H^+]$

There are two general processes that cause a reduction in $[H^+]$ of body fluids: (a) loss of volatile acid (CO_2) and (b) addition of base.

Loss of Volatile Acid Loss of CO_2 (for example, decreased arterial $[CO_2]$ due to hyperventilation) will lead to a decrease in the velocity of dissociation; thus, $V_{a_1} > V_{d_1}$ (Equation 5-29). During the period of disequilibrium $[H^+]$ and $[HCO_3^-]$ will both decrease. The end result after restoration of equilibrium at a reduced arterial P_{CO_2} is an increased pH (decreased $[H^+]$) and a decreased $[HCO_3^-]$.

Addition of Base The concentration of HCO_3^- will increase and $[H^+]$ will decrease when bases are added to ECF. If the salt, Na^+B^-, is added to ECF, it can be seen from Equation 5-29 that V_{a_2} will increase so that the concentration of HB increases while $[H^+]$ decreases. Because there is a decrease in $[H^+]$, V_{a_1} is slowed and more HCO_3^- is formed from dissociation of H_2CO_3. After equilibrium is restored, the $[H^+]$ is decreased, $[HCO_3^-]$ is increased, and $[B^-]$ is increased. It should be noted that the final effects of addition of $Na^+HCO_3^-$ are the same as those obtained from addition of Na^+B^-. Table 5-2 summarizes the preceding discussion.

Table 5-2 Summary of Changes in $[H^+]$, pH, and $[HCO_3^-]$ Occurring After Addition and Loss of Acids and Bases to ECF

	$\Delta[H^+]$	ΔpH	$\Delta[HCO_3^-]$
Addition of CO_2	↑	↓	↑
Addition of fixed acid	↑	↓	↓
Loss of HCO_3^-	↑	↓	↓
Loss of CO_2	↓	↑	↓
Addition of base	↓	↑	↑

Table 5-3 Changes in pH and $[HCO_3^-]$ Following Addition of CO_2 to ISF in Vitro

P_{CO_2}, torr	pH	$[HCO_3^-]$, mEq / L
20	7.70	24
40	7.40	24
80	7.10	24

BUFFERING OF VOLATILE ACIDS IN VITRO

Buffering of volatile acid is more easily understood if the chemical reactions in a given fluid such as ISF are first presented without the complicating influence of exchanges between body fluid compartments. It would be impossible to add CO_2 to the ISF compartment without at the same time adding CO_2 to the blood and ICF. Furthermore, as discussed in the chapter on blood-gas transport, CO_2 and H_2O combine in the red cells to form HCO_3^-, which then diffuses into plasma and ISF. For this reason the discussion in this section focuses on buffering of three body fluids (ISF, plasma, and blood) in vitro, that is, in glass or an artificial environment. In the next section buffering of volatile acid is described for the in vivo setting, where the influence of exchanges between body fluid compartments plays a role.

ISF It is instructive to consider the changes in pH and $[HCO_3^-]$ that occur when CO_2 is added to a fluid in which the only buffer system present in sufficiently high concentrations for significant buffering is the HCO_3^- system. Therefore, CO_2 will be added to ISF in vitro by changing P_{CO_2}. To do this, CO_2 is added to the fluid using a tonometer, a device in which fluids may be equilibrated with various gas mixtures.

Adding CO_2 to ISF in vitro is equivalent to adding CO_2 to ISF in vivo without allowing interactions with buffers of other compartments. In the following example (Table 5-3) it will be assumed that the ISF has a pH of 7.40, a $[HCO_3^-] = 24$ mEq/L, and a $P_{CO_2} = 40$ torr. Volatile acid (CO_2) is first added by increasing the P_{CO_2} from 40 to 80 torr and then removed by

FIGURE 5-2
CO_2 titration curves for ISF, plasma, and whole blood.

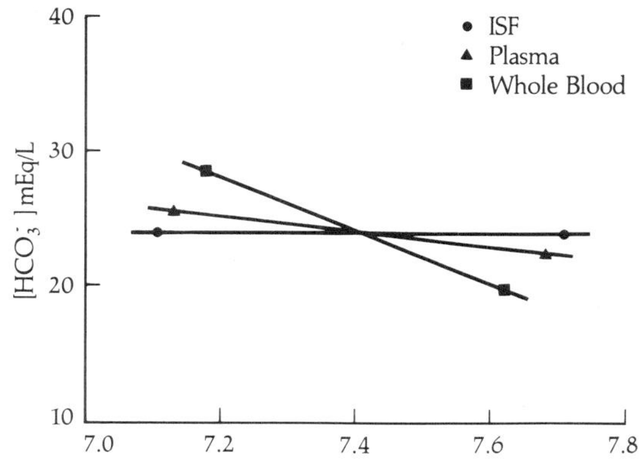

decreasing P_{CO_2} to 20 torr. The pH will be measured when the P_{CO_2} is at 80 and 20 torr. The $[HCO_3^-]$ can be calculated from Equation 5-28 or determined using Figure 5-1. The results of such a hypothetical experiment are presented in Table 5-3 and Figure 5-2.

It is apparent from inspection of Figure 5-2 and Table 5-3 that there is no measurable change in $[HCO_3^-]$ even though the P_{CO_2} changed fourfold. At first glance this appears to be a direct contradiction of what is stated in the previous section and Table 5-2. The fact is, $[HCO_3^-]$ did change as predicted. The predictions made in Table 5-1 are based on Equation 5-29. The following reaction describes events in ISF:

$$CO_2 + H_2O \rightleftharpoons H_2CO_3 \rightleftharpoons H^+ + HCO_3^- \quad (5\text{-}30)$$

Note that there is no cross-reaction with other weak acid-base (buffer) systems. In fact, when P_{CO_2} was doubled from 40 to 80 torr, the $[H^+]$ also doubled from 40×10^{-9} Eq/L to 80×10^{-9} Eq/L. For every *new* H^+ formed, a *new* HCO_3^- was formed. Thus, 40×10^{-9} Eq of H^+ were added to each liter of ISF and 40×10^{-9} Eq of HCO_3^- were also added to each liter. By how much did the $[HCO_3^-]$ change? The changes are summarized in Table 5-4.

The $[HCO_3^-]$ did change, but not significantly. There are two reasons the change is undetectable. First, HCO_3^- is found in much higher concentrations in ISF than H^+, hence, even if $[H^+]$ doubles when P_{CO_2} doubles the effect on $[HCO_3^-]$ is not measurable. Second, there is *no* buffering of volatile acid in ISF. Volatile acid was added to a solution that contained only the volatile acid buffer pair (CO_2-HCO_3^-). Adding CO_2 to ISF is analogous to adding acetic acid to vinegar. Another source of base is needed to produce buffering. Because ISF lacks significant concentrations of other bases, volatile acid cannot be buffered. Note the very flat slope to the buffer line of ISF in Figure 5-2. There is no significant change in $[HCO_3^-]$ and, therefore, no buffering.

Plasma When the experiment on ISF is repeated for plasma, the following results will be obtained (Table 5-5 and Figure 5-2).

Table 5-4 Changes in $[HCO_3^-]$ of ISF Produced by Changing P_{CO_2}

	$P_{CO_2} = 80$ torr	$P_{CO_2} = 20$ torr
Initial $[HCO_3^-]$, Eq / L	0.024000000	0.024000000
Change in $[HCO_3^-]$	+0.000000040	−0.000000040
Final $[HCO_3^-]$	0.024000040	0.0239999960

Table 5-5 Changes in pH and $[HCO_3^-]$ Obtained When CO_2 is Added to Plasma in Vitro

P_{CO_2}, torr	pH	$[HCO_3^-]$, mEq / L
20	7.675	23
40	7.400	24
80	7.125	25

When plasma is used instead of ISF, the $[HCO_3^-]$ changes significantly when CO_2 is added or removed. The reactions in plasma are

$$\begin{array}{c} \\ CO_2 + H_2O \rightleftharpoons H_2CO_3 \rightleftharpoons \end{array} \begin{array}{c} HPr \\ \upharpoonleft\downharpoonright \\ H^+ \\ + \\ Pr^- \end{array} \begin{array}{c} \\ + HCO_3^- \end{array} \tag{5-31}$$

where HPr and Pr^- represent the protein–proteinate buffer pair in plasma. The reactions can be divided arbitrarily into two steps. First consider addition of CO_2 with its consequent increase in $[H^+]$, just as occurred in ISF. The increased $[H^+]$ causes an increase in the velocity of association of Pr^- and H^+, thereby removing *some* of the additional H^+ ions that were produced when CO_2 was added. Second, because of the buffering by Pr^-, $[H^+]$ is returned toward its initial value, thereby causing more carbonic acid to dissociate, producing more HCO_3^- as well as H^+. As a result, both $[H^+]$ and $[HCO_3^-]$ increase measurably. The significant increase in $[HCO_3^-]$ is due to the buffering of carbonic acid by the plasma protein buffer system.

When P_{CO_2} is decreased, $[H_2CO_3]$ decreases and, therefore, the rate of association of H^+ and HCO_3^- will be greater than dissociation of carbonic acid. Hence, $[H^+]$ will decrease. Because of the reduced $[H^+]$, the rate of association of Pr^- and H^+ will also decrease. As a result, *some* of the H^+ lost will be replaced from dissociation of HPr. The replacement of H^+ causes more HCO_3^- to associate with H^+ until, at equilibrium, both $[H^+]$ and $[HCO_3^-]$ are measurably decreased. Again, the significant change in $[HCO_3^-]$ is due to the buffering by plasma proteins.

Whole Blood In whole blood, both plasma proteins and hemoglobin (Hb) buffers are available for buffering added carbonic acid. Again, volatile acid

Table 5-6 Changes in pH and [HCO_3^-] Produced When CO_2 Is Added to Whole Blood in Vitro

P_{CO_2}, torr	pH	[HCO_3^-], mEq / L
20	7.62	20
40	7.40	24
80	7.17	28

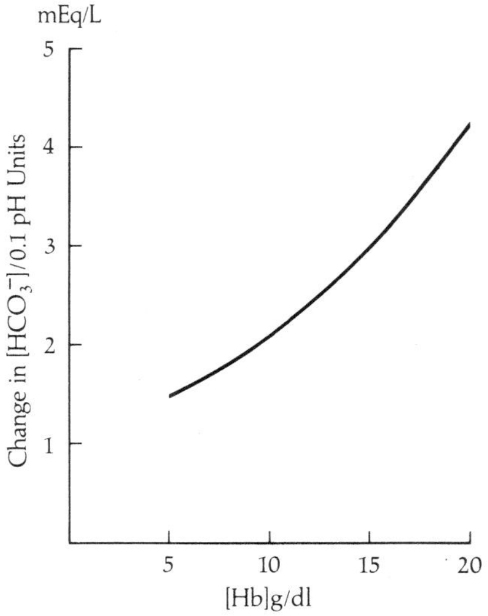

FIGURE 5-3
Change in [HCO_3^-] / 0.1 pH units as a function of hemoglobin concentration. This curve shows the change in slope of the CO_2 titration curve as a function of hemoglobin concentration. The more protein (including hemoglobin) buffer present in solution when CO_2 is added, the greater will be the corresponding change in [HCO_3^-] for a given pH change. The effect of protein concentration is also shown in Figure 5-2, where the slope of the CO_2 buffer curve increases with increasing concentration of protein in the solution being titrated. Plasma has more protein than ISF and whole blood has more protein than either plasma or ISF alone. The protein, thus, provides the buffer for H^+ dissociated from carbonic acid as shown in Equation 5-31.

concentrations can be changed by increasing and reducing the P_{CO_2}. The results from analysis of the blood at different values of P_{CO_2} are shown in Table 5-6 and Figure 5-2.

The same arguments that were used for plasma apply to buffering in whole blood except that the total concentration of protein buffer in whole blood (plasma proteins plus hemoglobin) is greater than plasma proteins alone, thus, when equilibrium is restored after addition or removal of CO_2, the change in [HCO_3^-] will be more pronounced.

Summary of Concepts Obtained from in Vitro Studies Two very important conclusions can be drawn from in vitro experiments. First, only nonvolatile buffer systems, that is, buffers other than HCO_3^-, can buffer volatile acid when CO_2 is added to or removed from body fluids. These nonvolatile buffers include plasma proteins, hemoglobin, and, as will be shown in the following section, intracellular phosphate and proteins. Second, the more concentrated the nonvolatile buffer, the more [HCO_3^-] changes and the less pH changes when CO_2 is added. This can be seen in the difference in slopes of the curves obtained for plasma and whole blood. This effect occurs not because

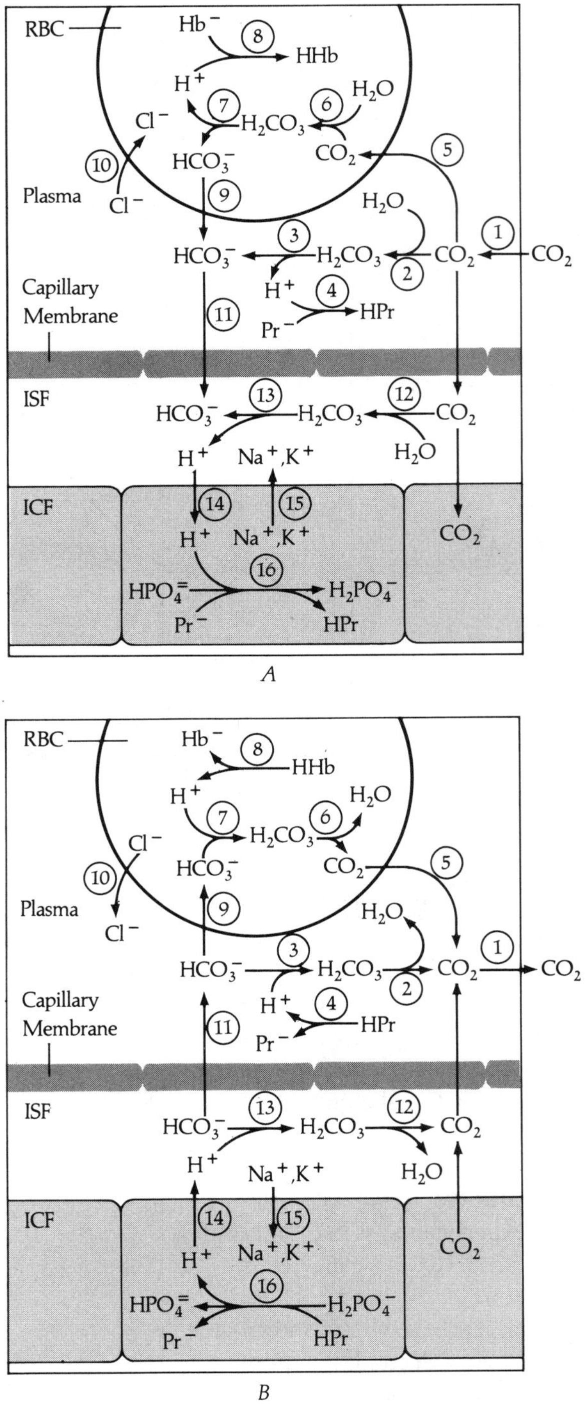

FIGURE 5-4 Model for addition of CO_2 to (*A*) and loss of CO_2 from (*B*) body fluid compartments. The numbers are given to identify specific reactions. Note that buffering of carbonic acid in the ECF occurs only in the plasma compartment (reaction 4). The bicarbonate produced in red blood cells from added CO_2 diffuses throughout the ECF (reactions 9 and 11 in *A*). There is also a significant exchange of K^+ and Na^+ for H^+ across cell membranes (reactions 14 and 15). This exchange is more pronounced with addition of CO_2 than with loss of CO_2 (see Figure 5-10).

hemoglobin is a different protein from plasma proteins but rather because whole blood has a greater concentration of total protein (plasma proteins plus hemoglobin) than plasma alone. The effect of protein concentration is also shown in Figure 5-3, where different samples of blood containing different concentrations of hemoglobin were titrated with volatile acid.

BUFFERING OF VOLATILE ACIDS IN VIVO

All of the principles discussed for buffering in vitro apply to buffering in vivo. The major difference between the two kinds of buffering is that electrolytes in plasma equilibrate with ISF electrolytes in vivo. Also, H^+ can exchange for Na^+ and K^+ across cell membranes between ISF and ICF. The different reactions and exchanges that occur as a result of addition of volatile acid in vivo are shown in Figure 5-4. It is important to remember that all the changes occurring in the different compartments are not measured by blood-gas analysis. Only changes in the plasma compartment can be readily measured by blood-gas analysis. Reactions 1 through 10 in Figure 5-4 are identical to those described for in vitro buffering of blood. However, in vivo, HCO_3^- diffuses out of the plasma compartment into the ISF (reaction 11, Figure 5-4). The effect of this diffusion is that $[HCO_3^-]$ of plasma does not increase as much in vivo for a given increment in P_{CO_2} as it does in vitro. In effect the HCO_3^- produced from dissociation of H_2CO_3 is distributed to a much larger volume in vivo (blood water plus ISF) whereas in vitro the HCO_3^- that is generated remains in the smaller volume of blood water.

Reaction 14 in Figure 5-4 represents exchange of H^+ for intracellular Na^+ and K^+ (reaction 15). This exchange between ISF and ICF has the opposite effect on $[HCO_3^-]$ of plasma from that shown in reaction 11. In effect, reaction 14 reduces $[H^+]$ of ISF, and thus, more dissociation of carbonic acid will occur (reaction 13). The result is an increase in $[HCO_3^-]$ of ISF and plasma. In

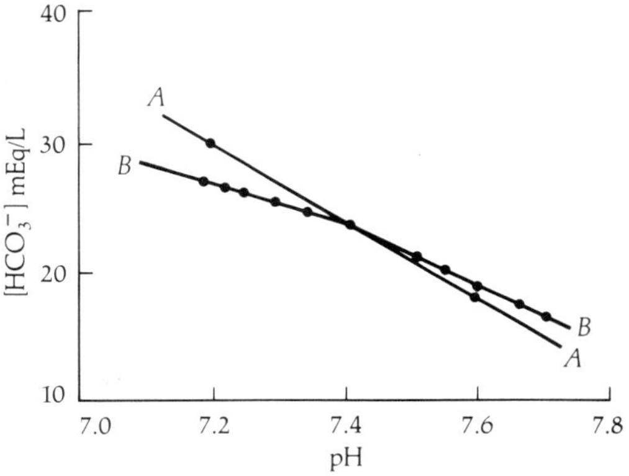

FIGURE 5-5
The in vitro (*A*) and in vivo (*B*) CO_2 titration curves. The curves were determined assuming the hemoglobin concentration was 15 g / dl.

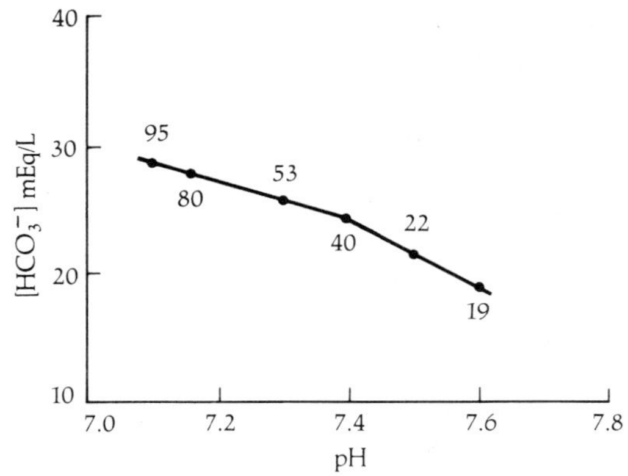

FIGURE 5-6

Values of P_{CO_2} along the in vivo CO_2 titration (or buffer) curve. Hemoglobin concentration was assumed to be 15 g / dl.

Source: Modified from "Basic mechanisms involved in acid-base homeostasis," by J. L. Keyes, *Heart and Lung,* 1976, 5:239–246. Reprinted by permission of The C. V. Mosby Company.

practice, the effects of reaction 11 predominate over those of reaction 14 so that in vivo the final steady-state value for [HCO_3^-] after addition of CO_2 is less than that which would be found in vitro.

From the preceding discussion it may be concluded that the slope of the CO_2 buffer (titration) curve in vivo is less than that found in vitro. Both the in vitro and in vivo curves are shown in Figure 5-5. The slope of the in vitro curve is -30 mEq HCO_3^-/L · pH unit (or -30 slykes). The in vivo curve is not linear. The slope from pH 6.9 to 7.4 averages -15 slykes whereas that from 7.4 to 7.8 averages -25 slykes. The steeper slope in the alkaline region is related to production of lactic acid that accompanies hypocapnia.

It is important to recognize that different points along the CO_2 buffer curve represent different values of P_{CO_2}. For example, at pH 7.4 the P_{CO_2} = 40 torr. At pH 7.1 the P_{CO_2} = 95 torr, and at pH 7.6 the P_{CO_2} = 22 torr. Other values for P_{CO_2} are shown in Figure 5-6.

BUFFERING OF NONVOLATILE ACIDS AND BASES IN BODY FLUIDS

Fixed acids and bases also accumulate in body fluids to produce acid-base disturbances. Changes that occur are described generally in the section titled "Reactions Involving the HCO_3^- Buffer System," using Equation 5-29, and are summarized in the second and fifth entries of Table 5-2. Now the changes will be presented graphically on the pH/[HCO_3^-] diagram.

Addition of Fixed Acids In Equation 5-29, let HB represent a fixed acid (for example, lactic acid) that is added to the blood from metabolism. It is assumed that the P_{CO_2} will remain constant during the addition of the acid. When HB is

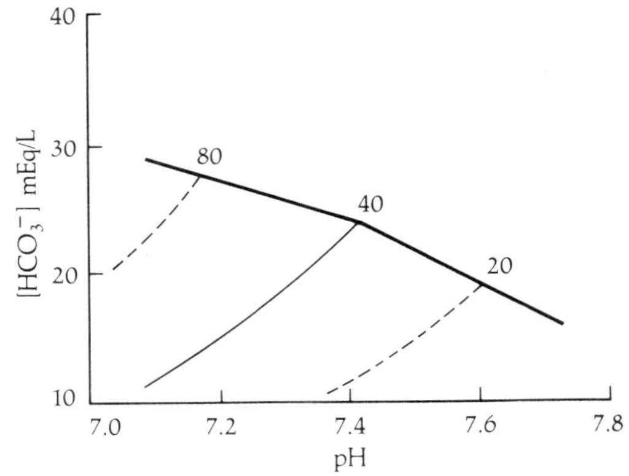

FIGURE 5-7
Changes in pH and bicarbonate concentration with addition of fixed acid at three different values of P_{CO_2}. When acid is added at a given (constant) P_{CO_2}, changes in the pH and bicarbonate concentration follow the P_{CO_2} isobar.

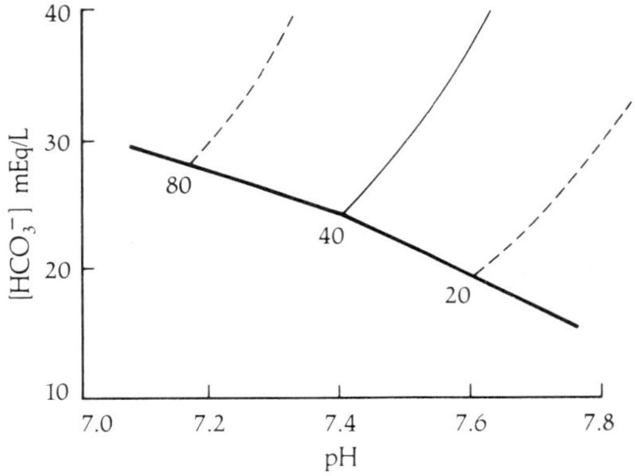

FIGURE 5-8
Changes in pH and bicarbonate concentration with addition of base at three different values of P_{CO_2}. When base is added at a given (constant) P_{CO_2} the increments in pH and bicarbonate concentration follow the P_{CO_2} isobar.

added, the $[H^+]$ of blood increases and $[HCO_3^-]$ decreases as described in the section "Reactions in Buffered Solutions." The decrement in $[HCO_3^-]$ can be determined from Equation 5-28 or the Siggaard-Andersen nomogram. The more acid added, the greater the change will be in both pH and $[HCO_3^-]$. The results of adding such an acid at a P_{CO_2} of 40 torr are shown by the solid curve in Figure 5-7. The acid could have been added at any P_{CO_2}, for example, 20 or 80 torr. Results of additions of fixed acid at these other values of P_{CO_2} are indicated by the dashed curves in Figure 5-7. In vivo buffering reactions to addition of fixed acids are shown in Figure 5-9.

Addition of Bases Addition of base occurs frequently in the clinical setting. Base may be added as lactate in lactated Ringer's solution or as HCO_3^- in excessive amounts by the kidneys. The general results of addition of base can be determined from Equation 5-29 and specific changes in $[HCO_3^-]$ from Equation 5-28. If a base is added at a constant P_{CO_2} of 40 torr, both pH and $[HCO_3^-]$ increase as shown in Figure 5-8. Again, any P_{CO_2} could have been chosen. The dashed curves in Figure 5-8 show the results of adding base at a P_{CO_2} of 20 and 80 torr.

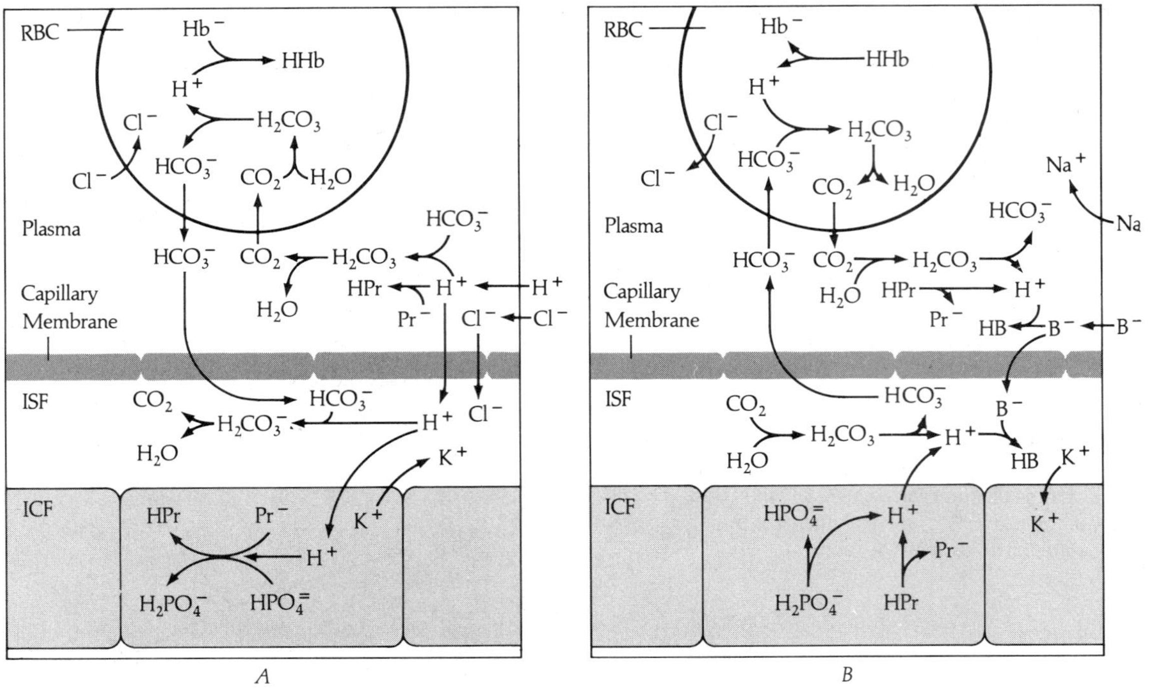

FIGURE 5-9 Addition of fixed acids (*A*) and base (*B*) to body fluid compartments. Percentages for each type of buffering reaction are shown in Figure 5-10. The buffering is assumed to occur at constant P_{CO_2} in these diagrams.

FIGURE 5-10 (*opposite*) Percent contribution of different processes to buffering. (*A*) Mechanisms of buffering of CO_2 in respiratory acidosis in the dog. (*B*) Mechanisms of buffering of CO_2 in respiratory alkalosis in the dog. $H^+ \rightleftharpoons Na^+$ and $H^+ \rightleftharpoons K^+$ are reactions 14 and 15 in Figure 5-4; $Cl^- \rightleftharpoons HCO_3^-$ are reactions 9 and 10 in Figure 5-4. Lactic acid (H^+, Lac^-) is formed in respiratory alkalosis and the H^+ from that acid contributes 35% of the buffering in this disorder. $HPr \rightleftharpoons Pr^- + H^+$ is reaction 4 in Figure 5-4. (*A* and *B* adapted from Giebisch, G., Berger, L., and Pitts, R. F.: *J. Clin. Invest.* 34:231, 1955.) (*C*) Mechanisms of buffering of strong acid infused intravenously in the dog. H^+, Cl^+ represents uptake of H^+ and Cl^- across cell membranes, which occurs primarily in the liver. (Adapted from Swan, R. C., and Pitts, R. F.: *J. Clin. Invest.* 34:205, 1955.) (*D*) Mechanisms of buffering of base infused intravenously in the dog. (Adapted from Swan, R. C., Axelrod, D. R., Seip, M., and Pitts, R. F.: *J. Clin. Invest.* 34: 1795, 1955.)

Reproduced with permission from *Physiology of the Kidney and Body Fluids,* 3rd Ed., by R. F. Pitts. Copyright © 1974 by Year Book Medical Publisher, Inc., Chicago. (Adapted originally from R. C. Swan, D. R. Axelrod, M. Seip, and R. F. Pitts, *Journal of Clinical Investigation,* 34:1795, 1955).

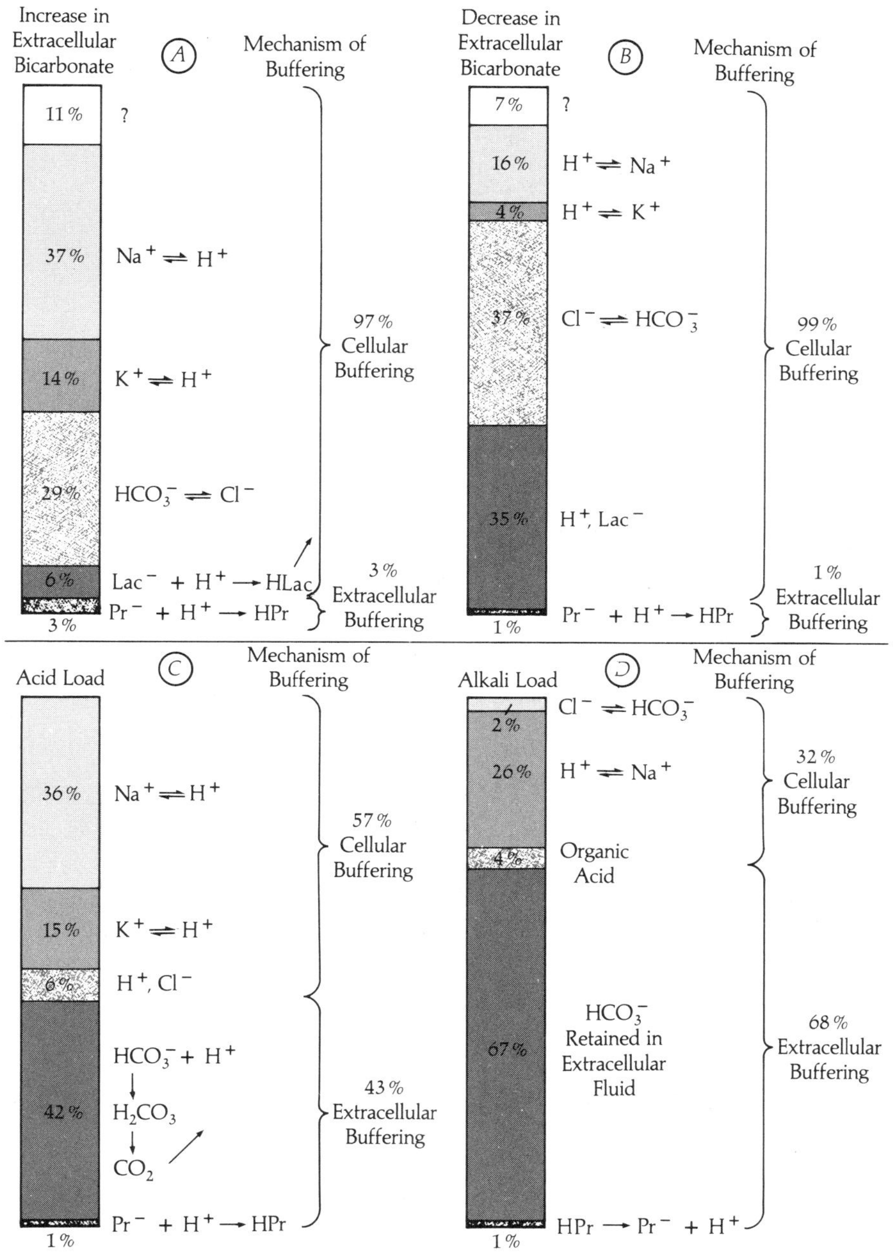

Increase in Extracellular Bicarbonate
A
Mechanism of Buffering
11%
?
37%
Na+ ⇌ H+
14%
K+ ⇌ H+
29%
HCO3− ⇌ Cl−
6%
Lac− + H+ → HLac
Pr− + H+ → HPr
3%
97% Cellular Buffering
3% Extracellular Buffering
Decrease in Extracellular Bicarbonate
B
Mechanism of Buffering
7%
?
16%
H+ ⇌ Na+
4%
H+ ⇌ K+
37%
Cl− ⇌ HCO3−
35%
H+, Lac−
Pr− + H+ → HPr
1%
99% Cellular Buffering
1% Extracellular Buffering
Acid Load
C
Mechanism of Buffering
36%
Na+ ⇌ H+
15%
K+ ⇌ H+
6%
H+, Cl−
42%
HCO3− + H+
H2CO3
CO2
Pr− + H+ → HPr
1%
57% Cellular Buffering
43% Extracellular Buffering
Alkali Load
D
Mechanism of Buffering
Cl− ⇌ HCO3−
2%
26%
H+ ⇌ Na+
4%
Organic Acid
67%
HCO3− Retained in Extracellular Fluid
HPr → Pr− + H+
1%
32% Cellular Buffering
68% Extracellular Buffering

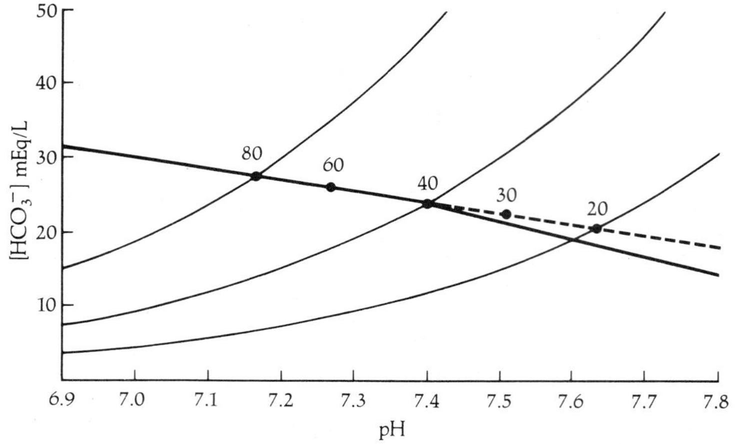

FIGURE 5-11 The pH / [HCO_3^-] diagram for the physiologic pH range. The P_{CO_2} isobars for 20, 40, and 80 torr are shown. Values for 30 and 60 torr are also included. The CO_2 buffer curve shown in this figure is an in vivo titration curve (hemoglobin concentration 15 g / dl). The solid portion of the curve represents the true in vivo titration curve throughout the physiologic pH range. The dashed portion of the curve represents the linear extrapolation from the acid range. If the extrapolation were used, the largest error in [HCO_3^-] would be about 2.5 to 3 mEq / L at a pH of 7.8. Ordinarily, in the alkaline range usually encountered clinically, errors of less magnitude would be found. As a result, many authors ignore the change in slope of the CO_2 buffer curve from pH 7.4 to 7.8. However, the solid curve shown above will be used in this text.

It is important to note that *when fixed acids and bases are added to body fluids, the HCO_3^- buffer system provides the initial buffering*. If the fixed acids and bases enter the red cell or the ICF compartments, other buffers such as hemoglobin and phosphate will contribute to buffering. However, in the extracellular fluid, the vast majority of buffering of added fixed acids and bases is from the HCO_3^- system. The sequence of events that occur is shown in Figure 5-9.

SUMMARY

Volatile Acid Addition and loss of CO_2 titrates blood along the in vivo buffer line (Figure 5-6). The contributions of the various compartments to buffering changes in volatile acid were determined in a series of elegant experiments in the 1950s under the direction of R. F. Pitts. The results he obtained are summarized in Figure 5-10.

Fixed Acids and Bases Addition of fixed acids titrates blood downward and to the left from the in vivo CO_2 buffer curve in the pH/[HCO_3^-] diagram. Addition of base titrates blood upward and to the right from the in vivo CO_2

buffer curve in the pH/[HCO_3^-] diagram (Figures 5-7 and 5-8). Pitts and his colleagues also determined the contributions of various compartments to buffering fixed acids and bases. These results are also shown in Figure 5-10.

Finally, the pH/[HCO_3^-] diagram permits clear separation of the primary causes of acid-base disturbances. The complete diagram with CO_2 buffer curve and P_{CO_2} isobars is shown in Figure 5-11.

CHAPTER 6

Pathophysiology of Acid-Base Disturbances

A PERSPECTIVE ON ACID-BASE DISTURBANCES

Acid-base disturbances are concomitant with and complications of disease processes but are not themselves diseases. Diseases of the lungs, kidneys, and cardiovascular system are common causes of acid-base disturbances. Treatment of an underlying disease process often corrects the associated acid-base disturbance. Nevertheless, if pH is shifted into an extreme and life-threatening range, it is necessary to treat both the acid-base disturbance and the causative disease processes. It should be remembered, however, that "overtreatment" can create another and often equally serious acid-base disturbance. Therefore, caution should be exercised so that the cure does not become worse than the disease itself.

The pathophysiology of acid-base disturbances includes the processes that bring them into being and the mechanisms of compensation that minimize the change in pH. *Compensation* of an acid-base disturbance is a set of processes generated by the respiratory system or kidneys in response to a change in pH. The term *compensation* does not refer to buffering or correction of the acid-base disturbance. Buffering is the immediate chemical response to change in $[H^+]$ of body fluids, whereas correction signifies a return of *all* acid-base variables to normal. Compensation is the result of the physiologic processes that attempt to return the pH to normal. These compensatory processes actually accentuate the abnormal values of P_{CO_2} and $[HCO_3^-]$ in blood.

Table 6-1 Normal Values for Arterial Blood-Gas Composition

Parameter	Range	Mean
pH	7.37 – 7.43	7.40
P_{CO_2}, torr	38 – 42	40
$[HCO_3^-]$, mEq / L	22 – 26	24
P_{O_2}, torr	80 – 100	*

**The value of P_{O_2} decreases with age. About 1 to 2 torr should be subtracted from 100 torr for each decade of age.*

Source: "Blood-Gas Analysis and the Assessment of Acid-Base Status," by J. L. Keyes, *Heart and Lung*, 1976, 5:247 – 255. Reprinted by permission of The C. V. Mosby Company.

Therefore, compensation is "expensive" in that when it occurs some other aspect of regulation is compromised.

There are two fundamental kinds of acid-base disturbances, acidosis and alkalosis. *Acidosis* is a process or set of processes that decreases the pH of body fluids to values less than normal. *Alkalosis* is a process or set of processes that increases pH of body fluids above normal.*

The processes of acidosis usually lead to acidemia whereas those of alkalosis usually cause alkalemia. However, there are types of acid-base disturbances in which pH remains within the normal range. In these latter situations, either the disturbance is well-compensated or both acidosis and alkalosis occur simultaneously. *Hence, normal pH by itself does not necessarily imply that an acid-base disturbance is not present. All acid-base parameters—pH, P_{CO_2}, and $[HCO_3^-]$—must be evaluated together to assess acid-base status.*

Acid-base disturbances originate from respiratory disorders, nonrespiratory (metabolic) disorders, or a combination of respiratory and metabolic disorders. For assessment purposes, it is useful to classify acid-base disturbances in terms of their origin. There are four categories: (1) respiratory, (2) metabolic, (3) combined, and (4) true mixed acid-base disturbances. Categories 3 and 4 involve specific combinations of 1 and 2. Compensation may occur in categories 1 and 2, but not in 3 and 4. The classification of acid-base disturbances is given in reference to normal blood-gas composition of arterial blood (Table 6-1). If the use of mixed venous blood for determining acid-base status becomes widespread, then definitions will need to be made in terms of mixed venous blood-gas composition as well.

*The definitions of acidosis and alkalosis vary somewhat from one textbook to another. Those stated above are based on the definitions given in the Report of the Ad Hoc Committee on Acid-Base Terminology in Current Concepts of Acid-Base Measurement, *Annals of the New York Academy of Sciences*, 133:257, 1966. "The terms 'acidosis' and 'alkalosis' describe abnormal processes or conditions which would cause a deviation of pH if there were no secondary changes in response to the primary etiologic factor." The terms acidosis and alkalosis "describe the overall process or condition without making such usage dependent upon deviation of pH per se."

Table 6-2 Causes of Respiratory Acid-Base Disturbances

Respiratory acidosis	Respiratory alkalosis
1. Airway obstruction	1. Mechanical ventilators
2. Pulmonary edema*	2. Pulmonary edema*
3. Pulmonary embolism*	3. Pulmonary embolism*
4. Pneumonia*	4. Hypoxemia
5. Neuromuscular diseases	5. CNS trauma and diseases*
6. Depression of the CNS	6. Drugs and hormones
7. Respiratory distress syndrome	7. Liver failure
8. Ascites	8. Anxiety
9. Obesity	9. Septicemia
10. Trauma to chest and lungs	10. Fever
11. Endocrine diseases*	11. Heat
12. Mechanical ventilators*	12. Pneumonia*

**Can cause either respiratory acidosis or alkalosis depending on circumstances.*

RESPIRATORY ACID-BASE DISTURBANCES

Respiratory acid-base disturbances originate as primary problems in respiratory function, either as diseases of the lung and/or chest wall, trauma to the lung and/or chest wall, or disorders in the control of ventilation. Regardless of the cause, the common denominator for respiratory acid-base disturbances is a change in the P_{CO_2} of arterial blood. A list of the more common causes of respiratory acid-base disturbances is shown in Table 6-2.

Respiratory Acidosis

In a steady state CO_2 production equals CO_2 excretion and the P_{CO_2} of arterial blood is constant (Chapter 4). This balance is frequently upset in disease. When the steady state changes, a respiratory acidosis may result. The fundamental cause of respiratory acidosis is hypercapnia. The P_{CO_2} of arterial blood can increase in three ways: (1) increased $F_{I_{CO_2}}$, (2) increased production of CO_2 ($\dot{V}_{CO_2}$), and (3) decreased excretion of CO_2 ($\dot{V}_E \times F_{E_{CO_2}}$).

The carbon dioxide concentration of clean air is essentially zero. When a person breathes a gas mixture containing CO_2 the P_{CO_2} of arterial blood will increase. This kind of gas mixture is used infrequently clinically but is sometimes used in the laboratory setting. The $F_{I_{CO_2}}$ may also increase when patients rebreathe expired air, as may occur with entrapment in a closed space or when rebreathing expired air from a bag. In any case, increased $F_{I_{CO_2}}$ is only rarely a cause of increased P_{CO_2} of arterial blood.

Increased production of CO_2 is rarely a cause of hypercapnia or respiratory acidosis. However, during parenteral hyperalimentation with solutions containing high concentrations of glucose, a patient may have episodes of hypercapnia from increased $\dot{V}_{CO_2}$.

The most common cause of respiratory acidosis is decreased excretion of CO_2. Hypercapnia is usually caused by decreased alveolar ventilation rate. The items in the left column of Table 6-2 all cause respiratory acidosis by means of decreased excretion of CO_2, that is, decreased alveolar ventilation rate. The details of the mechanisms for each item in Table 6-2 are beyond the scope of this text; however, they are explained in texts of respiratory pathophysiology and chest diseases. In each of these disease processes, the net effect is addition of CO_2 to body fluids.

The chemical buffering reactions that occur are shown in Figure 5-4. The general reaction is summarized in Equation 6-1:

$$CO_2 + H_20 \rightleftharpoons H_2CO_3 \rightleftharpoons \underset{\substack{+\\B^-}}{\overset{\substack{HB\\\upharpoonleft\downharpoonright}}{H^+}} + HCO_3^- \qquad (6\text{-}1)$$

When the P_{CO_2} of arterial blood increases, the $[H^+]$ and $[HCO_3^-]$ of blood also increase. The additional H^+ comes from the added CO_2 and buffering causes the measurable increase in HCO_3^- (Equation 6-1). The change in P_{CO_2} has a pronounced effect on secretion of H^+ and HCO_3^- generation in renal tubular cells. The increased P_{CO_2} in peritubular capillaries causes increased diffusion of CO_2 into renal tubular cells. This increased intracellular P_{CO_2} increases the rate of synthesis of H^+ and HCO_3^- by means of the reactions shown in Figures 4-9 and 4-10. The increased HCO_3^- production causes a rise in $[HCO_3^-]$ in plasma at the higher P_{CO_2}. In Figure 6-1 these processes are shown occurring in two steps. First, CO_2 is added to body fluids and pH shifts from the normal point (*N*) along the CO_2 titration curve to point 1. Buffering, but not compensation, has occurred at point 1; therefore, point 1 represents uncompensated respiratory acidosis. Increased renal production of HCO_3^- causes a shift from point 1 to point 2. This shift to point 2 represents the renal compensation of respiratory acidosis.

The compensation, that is, increased synthesis of HCO_3^- (through increased titratable acidity and NH_4^+ formation) and the resulting increased $[HCO_3^-]$ in arterial blood, reverses some of the effects of added CO_2 by returning pH toward normal. Note that, even with compensation, the P_{CO_2} remains at an increased value (for example, 80 torr) and the $[HCO_3^-]$ is increased over that produced by buffering. Thus, a compensated respiratory acidosis is characterized by an arterial pH less than 7.37, a P_{CO_2} greater than 42 torr, and a $[HCO_3^-]$ greater than 26 mEq/L (Table 6-3).

In Figure 6-1, changes are shown as a two-step process. This can occur with sudden sustained changes in P_{CO_2}. However, many respiratory diseases develop slowly over a period of months to years, and changes in P_{CO_2} and $[HCO_3^-]$ may be depicted as moving along a curve from the normal point to point 2.

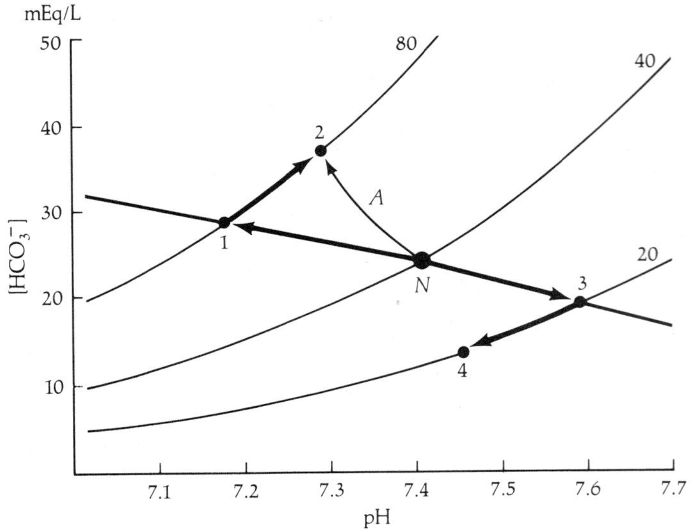

FIGURE 6-1
Respiratory acidosis and alkalosis. Acute addition of CO_2 causes an uncompensated respiratory acidosis (point 1). Point 2 shows compensation for respiratory acidosis. Curve *A* shows the development of compensated respiratory acidosis with chronic disease. Uncompensated respiratory alkalosis is shown at point 3 and with compensation at point 4. *N* is the normal point. *Source:* Modified from "Blood-Gas Analysis and the Assessment of Acid-Base Status," by J. L. Keyes, *Heart and Lung,* 1976, 5:247–255. Reprinted by permission of The C. V. Mosby Company.

Table 6-3 Arterial Blood-Gas Composition During Compensated Respiratory Acid-Base Disturbances

	Acidosis	Alkalosis
pH	< 7.37[†]	> 7.43[‡]
P_{CO_2}, torr	> 42	< 38
$[HCO_3^-]$, mEq / L	> 26*	< 22*

**Depending on pH. To be considered compensated, $[HCO_3^-]$ should be removed at least 2 to 3 mEq / L from the CO_2 buffer line (Figure 6-4).*

† *< signifies "less than."*

‡ *> signifies "greater than."*

Source: "Blood-Gas Analysis and the Assessment of Acid-Base Status," by J. L. Keyes, *Heart and Lung,* 1976, 5:247–255. Reprinted by permission of The C. V. Mosby Company.

Regardless of the time course, the increased P_{CO_2} stimulates renal compensation.

Respiratory Alkalosis

Respiratory alkalosis is caused by hypocapnia, which is produced by hyperventilation. Hyperventilation can be caused by fevers from sepsis, drugs including aspirin and progesterone, pyschogenic factors such as hysteria, vagal reflexes from lungs activated by pulmonary edema or pulmonary emboli, and overventilation with mechanical ventilators (Table 6-2).

Chemically, a respiratory alkalosis is produced by decreasing P_{CO_2} of body fluids, which in turn increases the pH and decreases $[HCO_3^-]$ (Equation 6-1).

This is illustrated as a shift along the CO_2 buffer curve from the normal point to point 3 in Figure 6-1. At point 3 the increase in pH has been buffered by addition of H^+ from proteins and hemoglobin, as shown by the reactions in Figure 5-4*B* and dissociation of HB in the vertical reaction in Equation 6-1. If the hypocapnia persists, then the kidneys reduce the rate at which H^+ is secreted because the stimulus from CO_2 is reduced. When the rate of H^+ secretion by renal tubular cells is reduced sufficiently, the HCO_3^- used in buffering fixed acids every day will not be totally replaced. In addition, some of the filtered HCO_3^- may be excreted because of the reduced H^+ secretion rate. In either case, the $[HCO_3^-]$ of plasma will decrease. The reduction in $[HCO_3^-]$ causes a shift down the new lower P_{CO_2} isobar from point 3 to point 4 (Figure 6-1). During compensation, decreasing the $[HCO_3^-]$ reverses some of the effects of the hypocapnia by causing more dissociation of carbonic acid (Equation 6-1), thereby increasing $[H^+]$. The net effect of compensation is to reduce pH toward normal while $[HCO_3^-]$ decreases even more than that from buffering.

Arterial blood-gas composition during compensated respiratory alkalosis will fall into the following ranges: pH greater than 7.43, P_{CO_2} less than 38 torr, and $[HCO_3^-]$ less than 22 mEq/L* (Table 6-3). It is not uncommon to have a patient present with an uncompensated respiratory alkalosis. This is frequently the case with psychogenic hyperventilation and fevers. In this circumstance, the $[HCO_3^-]$ will frequently vary between 23 and 22 mEq/L.

TIME COURSE FOR RENAL COMPENSATION

Renal compensation will normally not be completed in a short period of time. For respiratory acidosis, between 12 and 24 hours are needed for any measurable increase in $[HCO_3^-]$. The process of increased rate of synthesis begins very soon after the increment in P_{CO_2}; however, the HCO_3^- is distributed into the large ECF volume (ECF). A 50% increment in the rate of HCO_3^- synthesis would increase the $[HCO_3^-]$ by 2 mEq/L in 24 hours. On the other hand, compensation for a respiratory alkalosis can occur somewhat faster than for respiratory acidosis if excretion of HCO_3^- begins soon after the decrement in P_{CO_2}. Again, a few hours would be needed to produce any significant change in $[HCO_3^-]$.

RESPIRATORY ACID-BASE DISTURBANCES AND HYPOXIA

Hypoxemia can occur with either respiratory acidosis or alkalosis. In fact, hypoxemia is the primary cause of respiratory alkalosis associated with high

*Depending on pH. See the footnote in Table 6-3.

altitudes. When hypoventilation occurs and the inspired gas mixture is room air, hypoxemia will be present. Hence, a decreased P_{O_2} of arterial blood is frequently seen in respiratory acid-base disorders.

METABOLIC ACID-BASE DISTURBANCES

Metabolic acid-base disturbances are caused by accumulation of excess fixed acids and bases in body fluids. These disturbances usually arise from disorders of metabolism or systems other than the respiratory system. There is at least one important exception. When hypoventilation reduces arterial P_{O_2}, O_2 delivery is reduced to values at which anaerobic metabolism and then lactic acid production increase. Lactic acid is a fixed acid and when present in excess causes metabolic acidosis.

Metabolic Acidosis

Metabolic acidosis develops when excessive amounts of fixed acids are added to body fluids or when HCO_3^- is lost from body fluids. Normally fixed acids are added to body fluids at a rate of 50 to 100 mEq/day from metabolism. The kidneys generate HCO_3^- by forming titratable acidity and NH_4^+ (Chapter 4) to replace the normal loss of HCO_3^-. However, when fixed acid production increases, bicarbonate generation may not be able to keep pace with the loss, for example, with increased lactic acid formation from hypoxia and circulatory disturbances, ketoacidosis from diabetes, and loss of HCO_3^- from renal disease (Table 6-4). In renal disease the loss of HCO_3^- stems from decreased replacement of HCO_3^- used up in buffering of fixed acids produced daily from metabolism.

The hallmark of metabolic acidosis is a decrement in arterial $[HCO_3^-]$ accompanying a decreased pH. The common cause is increased production or addition of fixed acids to body fluids. The chemical reactions are shown in Equation 6-1. Assume that *HB* represents any fixed acid. Addition, or increasing the concentration of *HB*, increases the rate of dissociation of *HB*, and $[H^+]$ increases (vertical reaction). The increased hydrogen ion concentration in turn accelerates the association of HCO_3^- with H^+ to form carbonic acid (horizontal reaction). Because HCO_3^- is the primary buffer for added fixed acids, $[HCO_3^-]$ decreases when fixed acid concentration increases in plasma.

A primary reduction in $[HCO_3^-]$ as occurs in renal disease or diarrhea causes increased dissociation of carbonic acid, and hence $[H^+]$ increases. Whether $[H^+]$ changes by addition of fixed acids to body fluids or a primary decrement in $[HCO_3^-]$, the resulting changes, as displayed on the pH/$[HCO_3^-]$ diagram, are identical (Figure 6-2). Initially, the addition of the acid (or loss of HCO_3^-) causes a shift from the normal point, *N*, to point 1. The resulting increased

Table 6-4 Causes of Metabolic Acid-Base Disturbances

Metabolic Acidosis	Metabolic Alkalosis
1. Excess fixed acid production	1. Addition of bases
Lactic acid	Milk-alkali syndrome
Hypoxia	Administration of HCO_3^-
Decreased tissue perfusion	Lactated Ringer's solution
Salicylate intoxication	Citrated whole blood
Ketoacids	2. Generation of HCO_3^-
Diabetes	Gastric drainage
Starvation	Volume depletion
Ethanol intoxication	Diuretic therapy
2. Loss of HCO_3^-	Primary aldosteronism
Renal disease	Cushing's syndrome
Pancreatic drainage	
Diarrhea (especially in children)	
Acetazolamide	

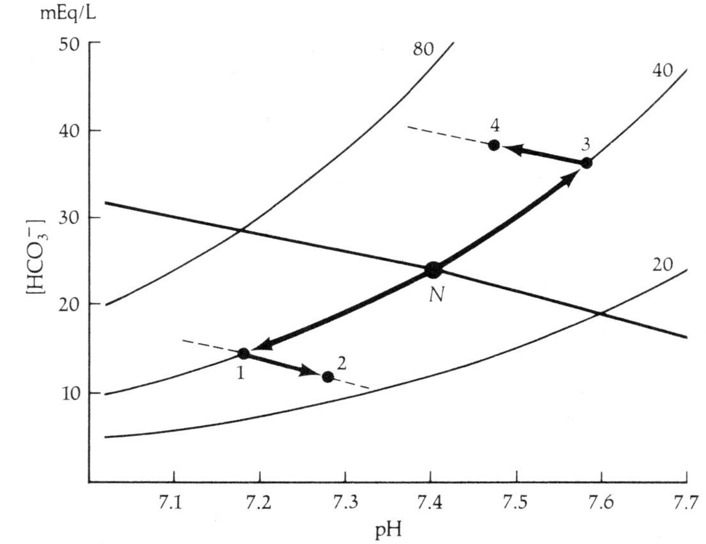

FIGURE 6-2
Metabolic acidosis and alkalosis. Point 1 shows the results of adding fixed acid at a constant P_{CO_2}, producing an uncompensated metabolic acidosis. Point 2 shows respiratory compensation. At point 3 base has been added which produces an uncompensated metabolic alkalosis, and point 4 represents respiratory compensation for the alkalosis. *Source:* Modified from "Blood-Gas Analysis and the Assessment of Acid-Base Status," by J. L. Keyes, *Heart and Lung,* 1976, 5:247–255. Reprinted by permission of The C. V. Mosby Company.

$[H^+]$ stimulates peripheral chemoreceptors (Chapter 4), and the ventilation rate increases within a few minutes of the onset of the primary disturbance. The increased ventilation rate is the compensatory response to addition of fixed acids; this is shown in Figure 6-2 as a shift from point 1 to point 2. This latter shift is parallel to the CO_2 buffer line. In effect, when fixed acids are added to body fluids, the CO_2 buffer line moves downward along the P_{CO_2} isobar parallel to its original position until it passes through point 1. When compensation causes a reduction in arterial P_{CO_2}, the change in P_{CO_2} must

Table 6-5 Arterial Blood-Gas Composition During Compensated Metabolic Acid-Base Disturbances

	Acidosis	Alkalosis
pH	< 7.37*	> 7.43†
P_{CO_2}, torr	< 38	> 42
$[HCO_3^-]$, mEq / L	< 22	> 26

* $<$ *signifies "less than."*

† $>$ *signifies "greater than."*

Source: "Blood-Gas Analysis and the Assessment of Acid-Base Status," by J. L. Keyes, *Heart and Lung*, 1976, 5:247–255. Reprinted by permission of The C. V. Mosby Company.

follow the CO_2 buffer line. Consequently, changes in P_{CO_2}, whether primary or compensatory, parallel the CO_2 buffer line.

The loss of CO_2 from hyperventilation decreases the $[H^+]$, and thus pH is returned toward normal. However, because CO_2 is lost in the compensatory process, $[HCO_3^-]$ decreases further from its value at point 1. The reduced P_{CO_2} will initially limit the compensatory response. However, as H^+ from the fixed acid crosses the blood-brain barrier, ventilation rate will be increased more.* During uncompensated metabolic acidosis (point 1) pH and $[HCO_3^-]$ will be decreased below normal values, but P_{CO_2} will remain at its initial value. With compensation, the P_{CO_2} of arterial blood will also decrease (Table 6-5).

Metabolic Alkalosis

Metabolic alkalosis is caused by two processes: (1) addition of bases to body fluids from exogenous sources and (2) generation of more HCO_3^- than is needed to replace that lost to buffering of fixed acids. Exogenous base may be added by ingestion or by infusion of HCO_3^-, lactate, or other bases. Increased generation of HCO_3^- can come from the kidneys in response to diuretic therapy, volume depletion, or Cl^- loss. The gastric mucosa also generates HCO_3^- as part of the process of secretion of HCl (Figure 6-3). When gastric acid secretion increases, as occurs during nasogastric suction or repeated episodes of vomiting, HCO_3^- synthesis by the stomach also increases (see Table 6-4).

When base is added to blood, the $[HCO_3^-]$ of plasma increases (see Figure 5-9). For example, if B^- in Equation 6-1 represents a base such as lactate, acetate, or citrate, the association reaction ($B^- + H^+ \rightarrow HB$) increases in rate and pH increases. As a result of the reduced $[H^+]$, more carbonic acid

*When blood $[HCO_3^-]$ decreases, there is a slow loss of HCO_3^- from brain ISF. The decreased $[HCO_3^-]$ in brain ISF causes dissociation of H_2CO_3 (horizontal reaction, Equation 6-1), which increases $[H^+]$. The increased $[H^+]$ stimulates central chemoreceptors. Therefore, both increased $[H^+]$ of brain ISF and loss of HCO_3^- cause increased ventilation.

FIGURE 6-3
Reactions involved in gastric secretion of acid. Note that the bicarbonate generated in the process is taken up into the blood. When secretion of acid increases, HCO_3^- generation also increases. C. A. is carbonic anhydrase.

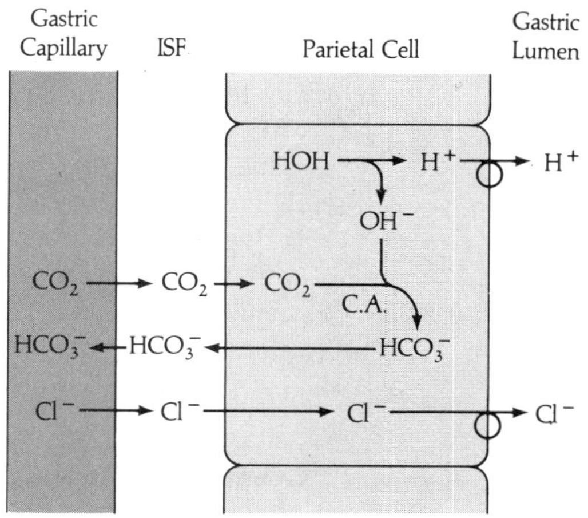

dissociates and $[HCO_3^-]$ increases (horizontal reaction in Equation 6-1). This process occurs initially at a constant P_{CO_2} in body fluids and is shown as a shift from the normal point (N) to point 3 in Figure 6-2. At point 3 the pH and $[HCO_3^-]$ are both increased, but P_{CO_2} has remained constant, therefore, only buffering has occurred.

The decreased $[H^+]$ in arterial blood caused by the alkalosis reduces stimulation to chemoreceptors, which ultimately produces a decrease in the ventilation rate. However, it is not the peripheral chemoreceptors that are responsible for the reduced ventilation rate. This is discussed in more detail later. The hypoventilation causes retention of CO_2 as described in Chapter 4. The net result is a shift from point 3 to point 4 in Figure 6-2. The shift occurs along the CO_2 buffer line. In metabolic alkalosis, the CO_2 buffer line moves upward from the normal point, N, parallel to its original position and passes through point 3. Consequently, when CO_2 is retained, the pH is compensated along the CO_2 buffer line to point 4. During compensated metabolic alkalosis all three acid-base variables are increased above normal (Table 6-5).

Hypercapnia is the compensatory response to metabolic alkalosis. The increased P_{CO_2} is a result of reduced stimulation to ventilation. The magnitude of the secondary hypercapnia is proportional to the change in $[HCO_3^-]$. For every milliequivalent-per-liter change in $[HCO_3^-]$ in metabolic alkalosis, the P_{CO_2} of arterial blood increases an average of 0.7 torr. This change in P_{CO_2} is about half that seen in metabolic acidosis, in which the compensatory change in P_{CO_2} averages about -1.2 torr for every milliequivalent-per-liter change in $[HCO_3^-]$.

Compensation for metabolic alkalosis is centrally mediated. Although peripheral chemoreceptors may have less than normal stimulation because of the increased pH, no change in arterial P_{CO_2} can occur until the $[HCO_3^-]$ of CSF increases. If the ventilation rate decreased immediately from the reduced stimulation of peripheral chemoreceptors, arterial P_{CO_2} would begin to increase.

Since CO_2 equilibrates quickly across the blood-brain barrier, the P_{CO_2} of the CSF would increase and central chemoreceptors would cause the ventilation rate to return to the higher initial value. However, with time the $[HCO_3^-]$ of the CSF will increase during metabolic alkalosis and the ventilation rate will be reduced. The rate-limiting factor for respiratory compensation of metabolic alkalosis is the process of equilibration of HCO_3^- with CSF.

Hypoventilation will cause arterial P_{O_2} to decrease. The reduction in P_{O_2} will limit the extent of compensation because hypoxia will stimulate peripheral chemoreceptors. The P_{O_2} of arterial blood has been reported to be reduced to values varying between 50 and 70 torr during compensated metabolic alkalosis.

COMPENSATION CURVES

In general, compensation for metabolic acid-base disturbances is mediated by changes in ventilation rate. Respiratory compensation usually returns pH about halfway to normal. Therefore, if metabolic acidosis decreases pH from 7.4 to 7.2, compensation will return it to approximately 7.3. Renal compensa-

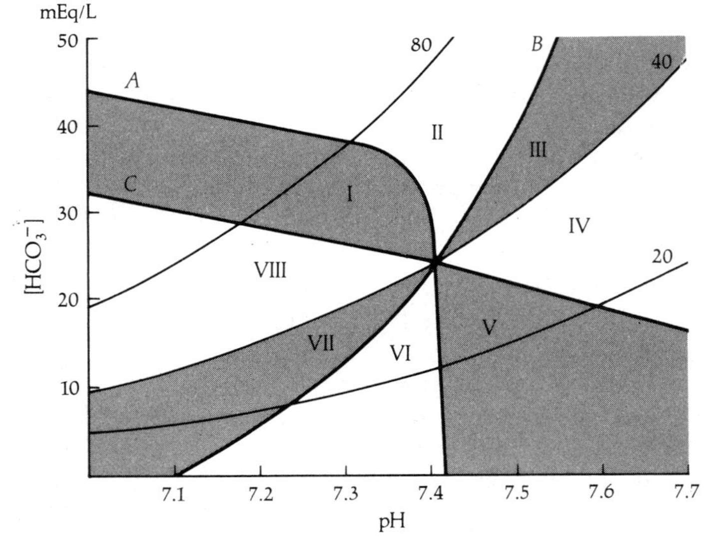

FIGURE 6-4 The pH / $[HCO_3^-]$ diagram with renal (*A*) and respiratory (*B*) compensation curves. These curves represent the maximum limits of compensation that would ordinarily be encountered. The Roman numerals show the different regions associated with each type of acid-base disturbance. Compensated disturbances are found in regions I, III, V, and VII. Complex disturbances (not compensated) are found in regions II, IV, VI, and VIII. *C* is the CO_2 buffer line. P_{CO_2} isobars at 20, 40, and 80 torr are also shown. (The compensation curves were determined by R. E. Swanson Ph.D. and are used in this figure with his permission.) *Source:* "Blood-Gas Analysis and the Assessment of Acid-Base Status," by J. L. Keyes, *Heart and Lung*, 1976, 5:247–255. Reprinted by permission of The C. V. Mosby Company.

tion is more complex. The degree of compensation depends on the magnitude of the initial change. Renal and respiratory compensation curves are shown in Figure 6-4.

These curves represent the maximum limit for compensation. The shaded areas represent regions of compensation and are classified as follows: area I, compensated respiratory acidosis; area III, compensated metabolic alkalosis; area V, compensated respiratory alkalosis; and area VII, compensated metabolic acidosis. When blood-gas composition lies outside of these areas of compensation, processes other than compensation must be occurring. Areas II, IV, VI, and VIII are regions of noncompensated acid-base disturbances. These are discussed in the next two sections.

A note of caution in interpretation is appropriate at this point. The compensation curves cannot be determined as precisely as the P_{CO_2} isobars. Consequently, the curves are really guidelines and should not be considered absolute demarcations. When a point determined from blood-gas analysis lies close to a compensation curve or even the CO_2 buffer line, the interpretation of acid-base status should be made with caution. History and repeated blood-gas analysis are frequently necessary to confirm an assessment. No competent clinician would decide on a specific therapeutic intervention if its outcome would be changed by inaccuracies inherent in the assessment process.

COMBINED METABOLIC AND RESPIRATORY ACID-BASE DISTURBANCES

Four of the regions in Figure 6-4 have not yet been classified. These regions comprise the mixed acid-base disturbances. There are two kinds of mixed disturbances: *combined** and true *mixed* acid-base disturbances. When both respiratory and metabolic processes combine and generate either acidosis or alkalosis, both respiratory and metabolic processes are involved in the production of the disturbance and no compensation can be generated.

Combined Acidosis

Combined acidosis occurs when respiratory and metabolic disorders occur together and both decrease the pH of body fluids. This problem can occur in patients who develop both respiratory and renal failure, marked hypoventilation or apnea, or a combination of reduced cardiac output coupled with hypoventilation.

*This type of acid-base disturbance is often called a mixed disturbance by some authors. However, it is useful to separate these types of mixed disturbances from true mixed disturbances to reduce the confusion.

Table 6-6 Blood-Gas Characteristics of Combined Acid-Base Disorders

1. Combined respiratory and metabolic acidosis:	
pH	less than 7.37
$PaCO_2$	greater than 42 torr
$[HCO_3^-]$	usually less than 26 mEq / L*
2. Combined respiratory and metabolic alkalosis:	
pH	greater than 7.43
$PaCO_2$	less than 38 torr
$[HCO_3^-]$	usually greater than 22 mEq / L†

*The value of $[HCO_3^-]$ will depend in part on the P_{CO_2} and the severity of the metabolic component. It is possible to have a combined acidosis in which pH = 7.1 at a P_{CO_2} of 90 torr and the $[HCO_3^-] = 27.1$ mEq / L. In combined acidosis the $[HCO_3^-]$ will be below the CO_2 buffer line on the pH / $[HCO_3^-]$ diagram.

† The value of $[HCO_3^-]$ will depend on the P_{CO_2} and the severity of the alkalosis. It is possible to have a combined alkalosis in which the pH = 7.65, and the $P_{CO_2} = 20$ torr and $[HCO_3^-] = 21.3$. In combined alkalosis the $[HCO_3^-]$ will be above the CO_2 buffer line on the pH / $[HCO_3^-]$ diagram.

The concentrations of both volatile and fixed acids increase in combined acidosis. Thus, in Equation 6-1 $[H^+]$ increases from two sources. The increased P_{CO_2} tends to increase $[HCO_3^-]$, but the effect of the fixed acid usually predominates and $[HCO_3^-]$ decreases. The reduction in $[HCO_3^-]$ is usually not as great as that observed for metabolic acidosis alone because of the opposing effects of CO_2 and fixed acids. In Figure 6-4, area VIII is the region that includes combined acidosis. The blood-gas composition of a patient with combined acidosis is described in Table 6-6.

Combined Alkalosis

A combined alkalosis occurs when both respiratory and metabolic disorders occur together and each alone would increase the pH of body fluids. In this type of acid-base disturbance hypocapnia is usually combined with an increase in $[HCO_3^-]$ of arterial blood. Examples include patients who are hyperventilated with a mechanical ventilator and are volume contracted from diuretic therapy or who are hyperventilated and are losing gastric secretions through nasogastric tubes. Arterial pH is typically increased and P_{CO_2} decreased. Bicarbonate concentration is usually slightly increased but may be normal. The $[HCO_3^-]$ is usually not increased as much as occurs with metabolic alkalosis alone because the decrement in P_{CO_2} tends to decrease $[HCO_3^-]$.

In Figure 6-4 area IV is the region of combined alkalosis. Arterial blood-gas composition for combined alkalosis is shown in Table 6-6. Both combined acidosis and alkalosis must be considered serious acid-base disturbances, because they frequently change pH rapidly to values that endanger life.

MIXED ACID-BASE DISTURBANCES

Mixed acid-base disturbances are produced when both respiratory and metabolic disorders occur simultaneously but one disorder generates alkalosis and the other acidosis. Under these conditions pH is frequently within the normal range or only slightly abnormal because the effects of the two disorders on pH are in opposite directions. The other blood-gas parameters, especially P_{CO_2} and $[HCO_3^-]$ are usually markedly abnormal. Unfortunately, a mixed disturbance is often erroneously interpreted as a well-compensated acid-base disturbance. Careful analysis of blood-gas data, however, will reveal that the point lies outside the four compensated regions (Figure 6-4), and, therefore, something other than a compensated disturbance exists. The two types of mixed disturbances are discussed in the next sections.

Mixed Metabolic Alkalosis and Respiratory Acidosis

This type of acid-base problem occurs when both respiratory acidosis and metabolic alkalosis occur simultaneously and both are primary disturbances, that is, neither is secondary to the other. Examples of this kind of acid-base disturbance include patients with COPD who lose gastric secretions from nasogastric suction and patients with COPD who are volume contracted because of diuretic therapy.

There is no compensation for this type of acid-base disturbance. The $[HCO_3^-]$ is greater than normal because two different processes increase the bicarbonate concentration. Hypercapnia, which is always present in this disorder, increases $[HCO_3^-]$, as shown by the horizontal reaction in Equation 6-1. Another source of HCO_3^- is also present and may include the kidneys, the stomach mucosa, or both. Occasionally the metabolic component is due to therapy from intravenous infusion of HCO_3^- or other solutions containing bases such as lactate or acetate. The pH does not usually change much unless one of the primary disorders predominates over the other. The increased P_{CO_2} from the respiratory component tends to decrease pH whereas the metabolic component tends to increase pH. The blood-gas composition of a patient with this type of mixed acid-base disturbance falls into area II in Figure 6-4.

Mixed Metabolic Acidosis and Respiratory Alkalosis

The second type of mixed acid-base disturbance occurs in patients who have metabolic acidosis and respiratory alkalosis concomitantly and neither is secondary to the other. Examples include patients who are mechanically hyperventilated and have a simultaneous lactic acidosis from circulatory failure or mechanically hyperventilated patients with renal failure.

In this type of acid-base disturbance, bicarbonate concentration of plasma is decreased from two causes. First, hyperventilation is always present and the resulting hypocapnia decreases [HCO_3^-] (Equation 6-1). Second, a metabolic component from addition of fixed acids (for example, lactic acid) or from loss of HCO_3^- also contributes to the decreased [HCO_3^-]. Because the effects on [H^+] oppose each other, the arterial pH will usually not deviate too far from normal unless one of the disorders predominates markedly over the other. The blood-gas composition of patients with this type of disorder will be found in region VI in Figure 6-4.

It should be remembered that management of patients with mixed acid-base disturbances can be difficult, especially when one of the components causing the disturbance begins to resolve. At that point the other component becomes predominant and pH can change quickly to extreme values. It is important, therefore, to monitor closely those patients with mixed acid-base disturbances. Often serial determinations of blood-gas composition are needed to properly assess and manage these patients. A summary of changes in all the different types of acid-base disturbances is presented in Table 6-7.

Table 6-7 Summary of Changes in Blood-Gas Composition from Normal Values during Acid-Base Disturbances

Type of disturbance	pH	P_{CO_2}	[HCO_3^-]	P_{O_2}*
Uncompensated disturbances				
Respiratory acidosis	↓	↑	± ↑	↓
Respiratory alkalosis	↑	↓	± ↓	Variable
Metabolic acidosis	↓	±†	↓	Variable
Metabolic alkalosis	↑	±	↑	Variable
Compensated disturbances				
Respiratory acidosis	↓	↑	↑	↓
Respiratory alkalosis	↑	↓	↓	Variable
Metabolic acidosis	↓	↓	↓	Variable
Metabolic alkalosis	↑	↑	↑	↓
Mixed disturbances				
Combined acidosis	↓	↑	± ↓	↓
Combined alkalosis	↑	↓	± ↑	Variable
R. acidosis + m. alkalosis	±	↑	↑	↓
M. acidosis + r. alkalosis	±	↓	↓	Variable

*P_{O_2} *is the least predictable of the variables. The arrows indicate the usual direction of change, but it should be kept in mind that change in* P_{O_2} *may be different from that indicated.*

† ± = *either no change or very slight change.*

CHANGES IN P_{O_2} DURING ALKALEMIA

The partial pressure of oxygen in arterial blood can vary widely in acid-base disturbances. It is frequently necessary to increase the $F_{I_{O_2}}$ or use mechanical ventilation, or both, to improve oxygenation of arterial blood. During alkalemia, however, an increase in arterial P_{O_2} may not be associated with improved tissue oxygenation. For example, a patient with a decreased arterial P_{O_2} and metabolic alkalosis may be mechanically ventilated to improve oxygenation of his or her arterial blood. However, if increasing ventilation rate by mechanical ventilation also reduces the P_{CO_2} of arterial blood below normal, then the attempt to improve arterial P_{O_2} will generate a combined alkalosis. The net result is that oxygenation of tissues may actually decrease while P_{O_2} of arterial blood increases. There are two reasons for the reduced oxygenation of systemic tissues. First, during alkalemia the increased pH increases the affinity of hemoglobin for O_2 (Bohr effect); thus, a lower tissue P_{O_2} is necessary to remove the same amount of O_2 from hemoglobin (see Figure 3-8). Second, positive pressure mechanical ventilation can cause a decreased cardiac output, which also reduces delivery of O_2 to systemic capillaries. Therefore, an increased arterial P_{O_2} reflects improved oxygenation of blood, but not necessarily improved tissue oxygenation. Consequently, when improved oxygenation of arterial blood is needed, care should be taken to avoid generating a concomitant alkalemia.

SUMMARY

The respiratory, cardiovascular, and urinary systems interact to maintain normal acid-base status. The interactions of these three systems maintain a normal balance or steady state for $[H^+]$, $[CO_2]$, and $[HCO_3^-]$ in body fluids. In a steady state, production or uptake of a substance is equal to metabolism and excretion of the substance. For example, more than 300 L of CO_2 are produced every day from metabolism. This amount of CO_2 is converted to approximately 15,000 mEq of carbonic acid per day. The carbonic acid is then dehydrated to CO_2 and H_2O and the CO_2 is excreted from the lungs at the same rate it is produced.

Bicarbonate concentration of extracellular fluid is kept constant by renal synthesis and excretion of HCO_3^-. Normally, 50 to 100 mEq of HCO_3^- are used up per day, and this amount is replaced daily by the kidneys. In the normal individual, the pH, P_{CO_2}, and $[HCO_3^-]$ are stable and within the normal range for these parameters. Such a person is said to be in acid-base balance or homeostasis, that is, a steady state.

Four processes are involved in maintaining acid-base balance: buffering, ventilation of the lungs, renal synthesis and excretion of HCO_3^-, and circulation of the blood.

Buffering is a chemical process in which hydrogen ions are added to or removed from solution when an alkali or acid is added to that solution. Changes in pH are thereby minimized. Buffering is the first line of defense against changes in acid-base balance, and it occurs in all body fluid compartments.

The respiratory system maintains alveolar ventilation at a rate sufficient to either increase or reduce CO_2 excretion to match production. The respiratory system can also increase and reduce ventilation rate to compensate or adjust for acid-base disturbances.

The kidneys generate (synthesize) HCO_3^- to replace that used in buffering. In addition, under the proper conditions they can excrete HCO_3^-. In health, extracellular HCO_3^- is used at a rate of about 50 to 100 mEq/day in buffering processes. The kidneys replace the HCO_3^- at the same rate. The kidneys can also increase or reduce HCO_3^- production when needed to compensate for acid-base disturbances. It should be noted that renal production of HCO_3^- is closely intertwined with regulation of ECF volume and Na^+ balance. Thus, alterations in fluid and electrolyte balance can, and do, affect acid-base balance.

Acid-base homeostasis requires adequate circulation of blood to supply O_2 to and remove waste products from the cellular environment. When the pH, P_{CO_2}, or $[HCO_3^-]$ of ISF changes, the value of these parameters also changes in plasma. Circulation of the blood to the appropriate receptors and organs (both peripheral and central chemoreceptors as well as kidneys) allows for detection, compensation, and correction of these changes. Thus, when circulatory failure occurs, serious acid-base disturbances can develop.

There are two major types of disturbances in acid-base homeostasis, acidosis and alkalosis. Each type is subdivided into categories depending on whether the disturbance is uncompensated, compensated, or mixed. Each type of acid-base disturbance is shown with its corresponding changes in blood-gas composition in Table 6-7.

Pathophysiology of Disturbances in Regulation of Volume and Osmolality of Body Fluids

CHAPTER 7

INTRODUCTION

Disturbances of fluid and electrolyte balance include disorders in the regulation of volume and osmolality of body fluids as well as changes in the normal concentrations of specific electrolytes. As was the case for acid-base disturbances, fluid and electrolyte disorders are usually complications of other disease processes and trauma. As such, these disturbances should not be thought of as isolated problems; rather they are part of the basic underlying pathophysiology of the diseases themselves. Because many diseases cause either loss or retention of solute and water, it is not surprising that fluid and electrolyte disturbances are common in the clinical setting.

There are several routes through which fluids and electrolytes may be gained or lost. Both normal and abnormal routes of intake and output are summarized in Figure 7-1. There are two kinds of input. Normally, input occurs through the gastrointestinal (GI) system when water, electrolytes, and food are ingested. These substances must then be absorbed into the blood to be delivered subsequently to the ISF and cells. On the other hand, fluids containing electrolytes and nutritive solutes can also be infused directly into the blood during parenteral administration of fluid.* Solutions may also be infused subcutaneously and intramuscularly, but the rate of infusion must, of neces-

*Parenteral, from the Greek *par* = beside and *enteron* = gut, refers to any route of input other than through the gastrointestinal pathway.

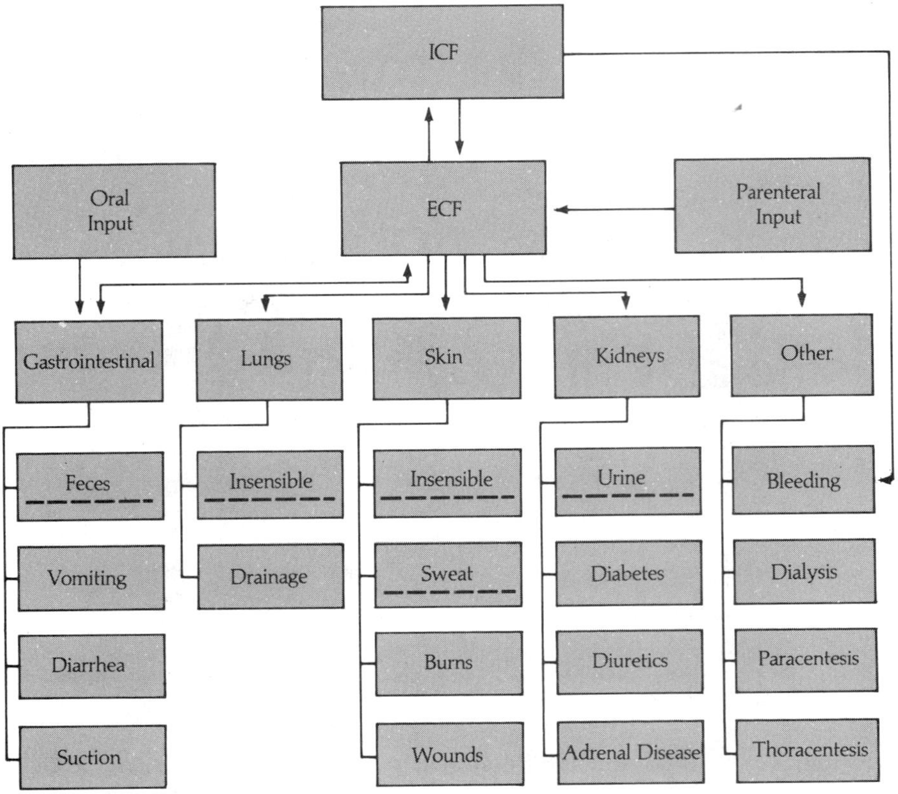

FIGURE 7-1 Input and output pathways. Oral intake is the normal input pathway into the body. Water and electrolytes enter the GI tract and are absorbed into the ECF. Normally, between 7 and 8 L of fluid are secreted into the GI tract daily and nearly all is reabsorbed. Of the volume entering the GI system per day, only about 150 ml leave in the feces. Other routes of *normal* excretion or loss (underlined with the dashed line) include insensible loss from the skin and lungs, sweat from the skin, and urine loss from the kidney. All other output pathways are abnormal routes of loss. All losses occur from the ECF, with bleeding or hemorrhage representing the only direct loss of ICF (as RBCs).

sity, be slower than when given by a direct intravenous route.

There are many more output pathways than input pathways. Those underlined in Figure 7-1 are normal pathways for output. The rest are considered abnormal routes or processes for loss of water and electrolytes. Loss can come from the upper gastrointestinal tract through vomiting or through drainage by means of a nasogastric tube. These tubes are commonly used following abdominal surgery. Excessive loss of fluid and electrolytes from the lower gastrointestinal tract can also occur from diarrhea. This type of loss may be very serious in young children and infants, because these individuals can quickly become dehydrated and develop circulatory insufficiency and shock. Very large diarrheal losses occur with cholera. In this latter case, parenteral fluid infusion is required to prevent hypovolemic shock and death.

Table 7-1 Classification of Disorders of Fluid Balance

I. Volume contraction (dehydration)	III. Edema
A. Isosmotic (isotonic)	A. Local
B. Hypoosmotic (hypotonic)	B. Regional
C. Hyperosmotic (hypertonic)	C. Generalized
II. Volume expansion	IV. Disturbances of osmolality
A. Isosmotic (isotonic)	A. Intake Disorders
B. Hypoosmotic (hypotonic)	B. Output disorders
C. Hyperosmotic (hypertonic)	

Insensible water loss through the lungs may increase significantly in hot, dry climates and in patients who are febrile. Normally, the skin helps contain or seal in water and electrolytes. There is, however, a normal insensible loss of water through the skin (see Chapter 2) and a loss of both water and electrolytes (chiefly sodium chloride) in sweat. These losses are increased in hot, arid environments. If the integument is damaged, such as occurs with wounds, skin ulcerations, burns, or abrasions, then large quantities of fluid and solute can be lost. In patients with extensive burns, for example, several liters of fluid may be needed just to replace losses in the first 24 hours.

The kidneys may excrete excess amounts of water and electrolytes in the urine in diseases such as adrenal insufficiency (Addison's disease), diabetes insipidus, or diabetes mellitus and with therapeutic use as well as abuse of diuretic drugs. Fluid loss following recovery from acute renal failure can also be severe enough to cause hypovolemia.

Finally, there are several other routes for fluid and electrolyte losses, including hemodialysis, peritoneal dialysis, thoracentesis, paracentesis, and of course hemorrhage. Note that all of these losses occur from the ECF through one of the major routes shown in Figure 7-1. There is no direct loss from the ICF unless tissue mass is lost, as may occur with trauma.

With all of these pathways for potentially abnormal losses in addition to normal fluid output, it would seem unlikely that volume expansion or overload could occur. However, there are several instances in which fluid excess is part of the pathology, such as heart failure, liver disease, renal failure, endocrine imbalance, and iatrogenic overinfusion. These examples are discussed in more detail in following sections. It should also be remembered that water and electrolyte balance are so tightly interwoven that disturbances in one will lead to imbalance in the other. Furthermore, even minor fluid disturbances, such as simple dehydration, can lead to development of acid-base disturbances. A classification of fluid balance disorders covered in subsequent sections is shown in Table 7-1.

VOLUME CONTRACTION

One of the most common types of fluid and electrolyte disturbances is volume contraction. This type of disorder is also called volume depletion or dehydration. Volume contraction leads to decreased blood volume (*hypovolemia*), which causes reduced perfusion of systemic tissues. Severe hypovolemia can cause prerenal failure, acute tubular necrosis, and hypovolemic shock. Therefore, some of the signs of volume contraction are those for circulatory insufficiency, that is, azotemia, oliguria, tachycardia, and decreased blood pressure. The decreased blood pressure causes increased activity of the renin-angiotensin control system, which is described in Chapter 2. As a result, in mild to moderate volume depletion, $[HCO_3^-]$ may increase, causing the generation of metabolic alkalosis. With more severe volume depletion, circulatory insufficiency causes tissue hypoxia and metabolic acidosis (see Chapter 6).

There are three principal types of volume depletion, as shown (Table 7-1). Each is named according to the osmolality of fluid remaining in the body. A model for each type is shown in Figure 7-2. In these models the body is portrayed as being composed of ECF and ICF compartments. This model, originally developed by Darrow and Yannet in 1935, is known as the Darrow–Yannet diagram. The height of the diagram is proportional to the osmolality of body fluids. Because cell membranes cannot maintain osmotic gradients between the ISF and ICF, the height of both compartments in the model will be the same at equilibrium. The width of each compartment is proportional to the volume of the compartment. Normally, the volume of the ICF is twice that of the ECF. The normal relationships are shown in part *A* of Figure 7-2. Changes in both osmolality and volume are shown by dashed lines.

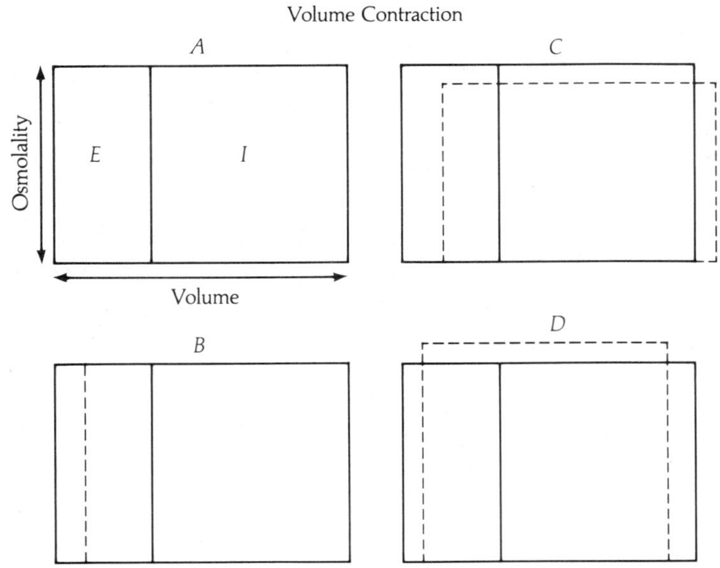

FIGURE 7-2 Types of volume contraction. (*A*) The normal relationships between extracellular fluid (*E*) and intracellular fluid (*I*). The width of each compartment is proportional to volume whereas the height is proportional to osmolality. The dashed lines show changes for each type of volume contraction; (*B*) isosmotic volume contraction; (*C*) hypoosmotic volume contraction; (*D*) hyperosmotic volume contraction.

Isosmotic Contraction

Isosmotic contraction of body fluids is shown in part *B* of Figure 7-2. In this situation, water and solute are both lost in proportions that maintain normal osmolality of the fluids that remain in the body. With isosmotic dehydration, the fluid lost comes from the ECF compartment alone. As portrayed in Figure 7-1, loss of any fluid from the body must come from the ECF initially through a route such as the kidneys, GI tract, lungs, skin, or dialysis. Fluid is not ordinarily lost directly from the ICF compartment. Therefore, if the fluid lost is isosmotic, there is no osmotic change that will initiate loss from the ICF. Recall from Chapter 2 that shifts from ICF to ECF and vice versa require changes in osmolality of the ECF. Because the ultimate result of isosmotic contraction is loss of volume from the ECF, without a change in osmolality, no loss of fluid can occur from the ICF.*

The most common causes of isosmotic dehydration are losses of fluid from the alimentary canal and hemorrhage. Approximately 8 L/day of fluid are secreted into the digestive tract. These fluids are essentially isosmotic to ISF. Hence, diarrhea, emesis, and gastric drainage can cause isosmotic dehydration. The most extreme example of GI fluid loss occurs with cholera. In this particular disease, several liters of isosmotic fluids are lost in a short period of time. The sodium concentration in the diarrheal fluid in cholera is essentially the same as that of plasma, therefore the sodium concentration of plasma does not change significantly. Hemorrhage is also an isosmotic loss of fluid from the body; however, some ICF is also lost with red cells. Technically, therefore, hemorrhage could be classified as isosmotic contraction of both ECF and ICF. When the volume lost from hemorrhage is small or moderate, only ECF (specifically plasma volume) needs to be replaced. When greater volumes are lost through bleeding, then red cells (ICF) as well as plasma volume are replaced.

Both hemorrhage and isosmotic contraction from GI losses produce signs and symptoms of volume depletion. In both, the decrement in capillary hydrostatic pressure reduces filtration across capillary walls into the ISF, and osmosis predominates (Figure 7-3). The net effect is that ISF is taken up into capillaries, helping restore *some* of the plasma volume that was lost. With hemorrhage, plasma proteins and red cells are lost. Consequently, when ISF is taken up into capillaries following hemorrhage, the remaining protein and red cell mass are diluted. The dilution causes hypoproteinemia and a reduced hematocrit (hemodilution). When isosmotic contraction occurs without blood loss, plasma proteins and red cells are first concentrated by the removal of water and electrolytes. Then, blood volume is *partially* restored by resorption of ISF. Therefore, when isosmotic volume contraction develops without blood

---

*This explanation is somewhat oversimplified. Osmolality changes can be followed by readjustment of water intake and output, so that final osmolality is normal but volume is contracted. It does not matter for purposes of classification what the intermediate steps were; it is only the final result that determines the type of fluid imbalance.

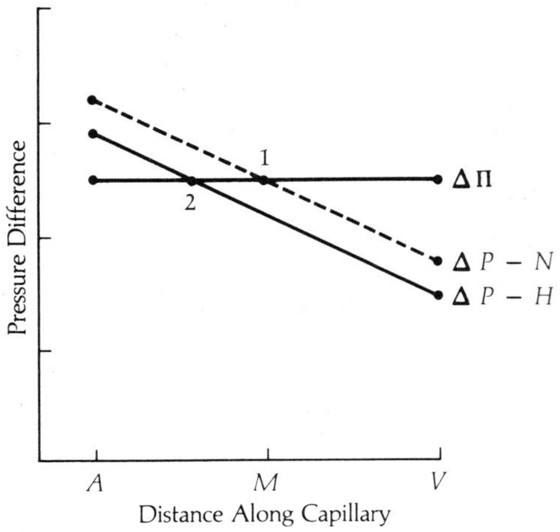

FIGURE 7-3
The effect of decreased hydrostatic pressure on net filtration pressure in the ideal capillary. The dashed line ($\Delta P - N$) represents the normal pattern of change in hydrostatic pressure difference along the length of the capillary. Note: $\Delta P = P_c - P_{isf}$, as described in Chapter 2, Equation 2-4. The solid line ($\Delta P - H$) represents a decreased hydrostatic pressure difference due to hemorrhage. A similar change would occur with marked venodilation. Point 1 at the midpoint (M) of the ideal capillary represents the equilibrium point where hydrostatic and oncotic ($\Delta\pi$) pressure differences are equal. Decreased hydrostatic pressure shifts the equilibrium to point 2, closer to the arteriolar (A) end of the capillary. As a result, net filtration pressure decreases and osmosis predominates. With hemorrhage, increased osmosis of ISF into the capillary temporarily helps compensate for the blood loss.

loss, both plasma protein concentration and hematocrit increase (hemoconcentration). The characteristics of isosmotic volume contraction are listed in Table 7-2.

Hypoosmotic Contraction

When loss of solute occurs without a proportional loss of water, that is, loss of hypertonic fluids, the body fluids remaining become hypoosmotic compared to normal, producing hypoosmotic contraction. The final result, which is shown in Figure 7-2*C*, may be considered as developing in two hypothetical steps. First, solute is lost from the ECF out of proportion to normal concentrations, creating a hypotonic extracellular fluid. Second, the hypotonic ECF causes water flow into the ICF by osmosis until osmotic equilibrium is restored. The end result is a decreased ECF volume, an increased ICF volume, and reduced osmolality of both compartments. Both hematocrit and plasma protein concentration are increased in this kind of dehydration. The increased hematocrit is caused by two processes: (1) osmotic swelling of red cells from hypotonicity of body fluids and (2) decreased plasma volume. The protein concentration in plasma will increase because of loss of plasma water.

Hypotonic contraction can be caused by problems such as adrenal insufficiency (Addison's disease), water replacement after sweating without adequate salt replacement, and diuretic therapy coupled with low sodium input. In Addison's disease there is a deficiency of all adrenal steroids because of atrophy or granulomatous destruction of the adrenal gland. The important missing steroid for purposes of this discussion is aldosterone. In the complete absence of aldosterone, about 2% of the total mass of sodium that is filtered daily plus accompanying anions are lost in the urine. Because of loss of this extra solute in the urine, additional water is needed for excretion, but the urine is hyperosmotic to ECF. The excretion of hyperosmotic urine leads to dilution

Table 7-2 Characteristics of Volume Depletion[*]

	Isosmotic contraction	Hypoosmotic contraction	Hyperosmotic contraction
Na^+	±	⇓	⇑
Cl^-	±	⇓	⇑
HCO_3^- [†]	⇑ [†]	⇑ [†]	⇑ [†]
Plasma protein[‡]	⇑	⇑	⇑
Hematocrit[‡]	⇑	⇑	± or ↑
BUN and creatinine[§]	⇑	⇑	⇑
Skin turgor	⇓	⇓	⇓

[*]*Key:* ⇑ *Increased compared to normal.*
⇓ *Decreased compared to normal.*
↑ *Slightly increased compared to normal.*
± *No change.*

[†]*The HCO_3^- concentration will depend on the severity of the volume loss. If the deficit is mild to moderate, volume contraction can cause an increased production and concentration of bicarbonate. If the deficit is more severe, circulatory insufficiency can cause the production of a metabolic (lactic) acidosis and the HCO_3^- concentration will decrease.*

[‡]*Hematocrit and plasma protein concentration can decrease if there is blood loss with the volume depletion. This is often the case with isosmotic contraction.*

[§]*BUN and creatinine concentration are inversely related to the GFR. If GFR decreases, then both BUN and creatinine concentrations increase.*

of the ECF. The volume loss also stimulates thirst and ADH release. The final result is hypotonic contraction of body fluids.

Profuse sweating can also cause hypotonic contraction of ECF when water is replaced without adequate replacement of salt. Sweat is hypotonic, and large losses initially cause hyperosmotic contraction of extracellular fluid. However, the volume depletion and hyperosmolality stimulate thirst, and water is usually ingested to help replace lost volume (see Figure 2-6).

Diuretic therapy with low sodium input occurs in several clinical situations. For example, patients with heart or vascular diseases are frequently treated with diuretic drugs to reduce ECF volume. Diuretics cause increased sodium and chloride excretion. The solutions used for fluid therapy in these patients typically contain little or no sodium salts. As a result, these patients are at high risk for developing hyponatremic volume contraction.

Hypotonic volume contraction can be of special concern when present in young children and infants. First, the extracellular fluid volume of a child is significantly less than that of an adult, and marked volume depletion can develop very quickly in these small patients. Frequently, the magnitude of fluid deficit is estimated from weight loss in the child. If the dehydration is isosmotic, then the weight loss provides a reasonable estimate of the deficit in ECF volume. However, if the dehydration is hypotonic, weight loss will be less than the decrease in ECF volume because some of the ECF shifts into the ICF compartment and does not appear as a change in weight. The danger is that the underestimated deficit in ECF volume will lead to an underestimate in the severity of the decrement in vascular volume, and hypovolemic shock could develop before adequate fluid therapy can be instituted.

Hyperosmotic Contraction

If water is lost out of proportion to solute loss, fluids remaining in body compartments become concentrated. Under these conditions the osmolality of both ICF and ECF will increase. Because the loss comes initially from ECF, concentration of the extracellular fluid causes osmosis from the ICF until osmotic equilibrium is restored. The net result is hyperosmolality of both ICF and ECF with loss of volume from both compartments (Figure 7-2*D*). Hematocrit may not increase very much with this type of dehydration because water moves out from red cells into the plasma in response to increased osmolality of ECF. Thus, when a blood sample is centrifuged to measure hematocrit, the volume of packed red cells will be reduced in proportion to plasma volume. The net effect is that hematocrit may be in the normal range. However, plasma protein concentration will be increased because of the dehydration. Other characteristics of hypertonic dehydration are shown in Table 7-2.

Hyperosmotic volume depletion is produced when individuals do not have access to fresh water, for example, as may occur when people travel in deserts or other arid places or are lost at sea. It is also encountered in patients who are unable to ingest water because of debilitation, coma, or certain gastrointestinal diseases. The simplest form of hypertonic dehydration is water loss without concomitant solute loss. For example, any time insensible water loss is not replaced, hyperosmotic volume depletion can develop. Hypertonic dehydration also occurs with solute loss when the lost volume is hypotonic to body fluids, such as occurs when sweat losses are not replaced.

VOLUME EXPANSION

The volume of fluid contained in body compartments may be increased under a variety of conditions. This volume expansion can be isosmotic, hypoosmotic, or hyperosmotic compared to normal extracellular fluid. As was the case for volume contraction, each type of expansion is named according to the concentration of the fluid remaining in the body after expansion has occurred (Figure 7-4).

Any solution added to body compartments gains entrance to the plasma and ISF from the digestive tract or by means of parenteral infusion. The final distribution of the added fluid depends on both its osmolality and its composition. If an isosmotic fluid is infused or absorbed from the alimentary canal into the extracellular space, there will be no change in the osmolality gradient

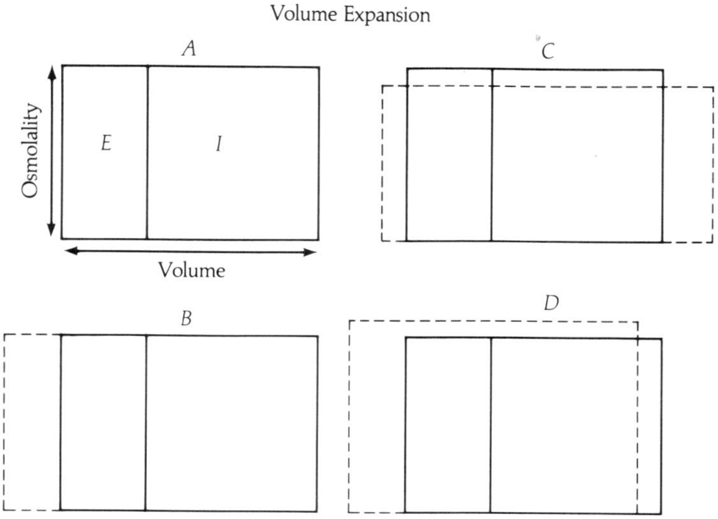

FIGURE 7-4 Types of volume expansion. (*A*) The normal relationships; (*B*) isosmotic volume expansion; (*C*) hypoosmotic volume expansion; and (*D*) hyperosmotic volume expansion.

between ISF and ICF. As a result, there will be no osmosis of fluid from the interstitial fluid into the intracellular fluid *unless* one or more of the added solutes crosses cell membranes. When an added solute enters the ICF, whether by diffusion or carrier-mediated transport, water will follow the solute by osmosis, thereby maintaining osmotic equilibrium between ICF and ECF. The more solute that enters the cells, the more water will enter and expand intracellular volume. Consequently, a solution composed of solutes that penetrate into the ICF, even if the solution is isosmotic, will cause expansion of both extracellular and intracellular compartments.

On the other hand, infusion of a solution that is isosmotic and contains only solutes that are restricted to the extracellular space will expand the extracellular fluid and not the ICF, because the added solutes do not accumulate in the intracellular space. Such solutions are said to be *isotonic*. Cells neither gain nor lose water when exposed to an isotonic solution. An isotonic solution is isosmotic and when given, expands the extracellular compartment but not the intracellular compartment. On the other hand, just because a solution is isosmotic does not mean that its volume of distribution is limited to the ECF.

Isotonic Expansion

Solutions commonly used for isotonic expansion of ECF are listed with their compositions in Table 7-3. Sodium is generally restricted to the extracellular compartment, and little accumulates across cell membranes in the ICF. Chlo-

Table 7-3 Examples of Commonly Used Parenteral Solutions*

Solutions	Na^+	K^+	Ca^{++}	Mg^{++}	NH_4^+	Cl^-	HCO_3^-	Glucose, g / L
Electrolyte								
Normal saline (0.9%)	154					154		
Ringer's lactate	130	4	2.7			109	28†	
3% Sodium chloride	513					513		
5% Sodium chloride	855					855		
0.9% Ammonium chloride					168	168		
Electrolytes with glucose								
5% Glucose and 0.225% sodium chloride	38.5					38.5		50
5% Glucose and 0.45% sodium chloride	77					77		50
5% Glucose and 0.9% sodium chloride	154					154		50

**Potassium chloride may also be added in amounts of 10, 20, 30, and 40 mEq / L. All values for electrolytes are expressed in milliequivalents per liter.*
†Converted from lactate in vivo as shown in Figure 7-5.

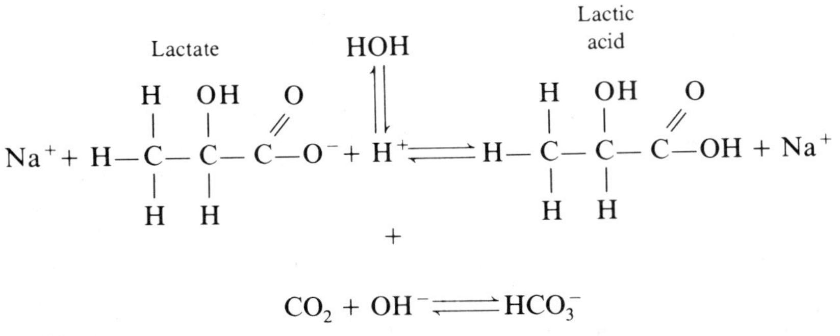

FIGURE 7-5 Metabolic conversion of lactate to HCO_3^-. In the body, sodium lactate is converted to sodium bicarbonate by the reactions shown. Actually, the base, lactate, combines with hydrogen ion from water to form the conjugate acid, lactic acid. The hydroxyl (OH^-) anion then combines with CO_2 to form the bicarbonate anion. Lactic acid is then metabolized to CO_2 and water. However, this conversion does not proceed in the absence of an adequate oxygen supply to metabolize lactic acid. The net result is that one bicarbonate anion is formed for every lactate metabolized.

ride, which is the major anion in most parenteral solutions, will remain in the ECF with sodium. The lactate will be converted to HCO_3^- as shown in the reaction in Figure 7-5. The HCO_3^- will either remain in the ECF or be used in a buffering reaction in which it will be replaced with the conjugate base from the acid that is buffered. In any event, the anions accompanying sodium in isotonic solutions also remain in the ECF. All those solutions will expand the ECF without a concomitant expansion of ICF (Figure 7-4*B*). This type of expansion will decrease the hematocrit and hemoglobin concentration as well as the total protein concentration of plasma. Isotonic expansion can occur as part of therapeutic intervention, iatrogenic overinfusion, and in diseases such as chronic heart failure, chronic renal failure, and chronic liver disease. Obstruction of blood vessels with tumors and clots may also lead to volume expansion. When the accumulation of fluid in the extracellular space is excessive, the individual is said to have edema or be edematous. Edema is discussed further in the section titled *Edema*.

Hypoosmotic Expansion

Fluid accumulation in the body that reduces osmolality below normal is called hypoosmotic or hypotonic expansion. Because the fluid that is accumulating is hypotonic compared to normal ECF, osmosis also occurs across cell membranes and the ICF compartment expands. Thus, with hypotonic expansion, both ECF and ICF compartments are increased in volume, and the osmolality of both is decreased (Figure 7-3*C*).

The hematocrit may not change much with mild hypotonic expansion, because water enters the red cells and causes swelling of the erythrocytes. As a result, the volume of red cells increases in proportion to the increased extracellular volume and the hematocrit does not change significantly. However, with marked expansion, hematocrit will decrease. Hypotonic expansion generally leads to hyponatremia and dilution of plasma protein concentration (hypoproteinemia).

Hypotonic expansion can occur with intake and retention of pure water and with infusion of hypotonic solutions containing solutes that cross cell membranes into the ICF. Adding pure water to ECF simply dilutes the fluid, and osmosis redistributes the water until osmotic equilibrium is restored. Infusion of solutions such as 5% glucose also produces hypotonic expansion. Initially, 5% glucose solutions are essentially isosmotic. However, the glucose enters cells and is metabolized, thereby removing the solute. Therefore, the net effect of infusion of 5% glucose solution on volume expansion is the same as for pure water. If sodium chloride is added to the glucose solution (Table 7-3), that amount of water needed to make the NaCl in the solution isosmotic with the prevailing conditions will be retained in the ECF with the sodium chloride. The more nonpermeant solute (in this case NaCl), the more water will be kept in the ECF. Other examples of hypotonic expansion are discussed in the section titled *Disorders of Osmolality of Body Fluids*.

Hyperosmotic Expansion

Hyperosmotic expansion occurs when hyperosmotic fluids* are added to body fluids. Infusion of these solutions initially raises the osmolality of the ECF, which causes osmosis of water from the ICF. When osmotic equilibrium is restored, the osmolality of both compartments will be increased over normal values. In addition, the volume of the ECF will be increased whereas that of the ICF will be decreased (Figure 7-3*D*).

The volume expansion of the ECF will dilute plasma proteins, causing hypoproteinemia. The hematocrit will decrease for two reasons. First, the added fluid plus the osmosis from ICF dilutes the red cell mass. Second, osmosis also occurs out of red cells, thereby reducing erythrocyte volume. Therefore, with hypertonic expansion, both hematocrit and plasma protein concentration will be reduced from normal.

The change in sodium concentration will depend on the type of solution used for expansion. If the expanding solution is hypertonic saline (Table 7-3), sodium concentration will increase. If no sodium is present in the infused hypertonic solution, sodium concentration will decrease until the solution is excreted or the solute in the solution is metabolized. For example, infusion of hypertonic glucose will initially increase osmolality of the ECF and cause a transient hypertonic expansion. After metabolism of the glucose, the expansion of body fluids would be hypotonic.

It should be noted that volume expansion, especially hypertonic, can lead to volume overload of the vascular space. This is a special concern in patients who are in acute or chronic renal failure. Ordinarily, the kidneys will excrete excess volume if expansion does not occur too rapidly. In renal failure, however, the normal mechanisms of excretion are not functioning properly, and hypervolemia often occurs.

EDEMA

Edema is the accumulation of an excessive amount of fluid in the interstitial space. Edema may be restricted or localized to a specific region or area such as occurs with an inflammatory response following minor tissue injury (minor burn, scratch, insect bite), it may be regional, or it may be generalized and distributed throughout the body such as occurs in cirrhotic liver disease and chronic heart failure.

*For the discussion in this section, the solutes making the infused fluid hypertonic are considered to be nonpermeable to the cell membrane.

Localized Edema: The Inflammatory Response

Whenever tissue is injured or invaded by foreign organisms, a complex set of processes localize the damage, destroy invading organisms, and initiate the first steps in repair of the tissues. The processes are called *the inflammatory response*. Inflammation is a nonspecific response of living tissue to injury. Characteristically, an inflamed region exhibits redness, is warm and painful to the touch, is swollen, and there is also a variable degree of loss of function. These characteristics are caused by three principal mediators: histamine, complement proteins, and kinins. A flowchart of the pathways and effects of the three mediators is shown in Figure 7-6. Redness and increased temperature of an inflamed area are caused by vasodilation, which increases blood flow and volume of blood (hyperemia) in the region. Kinins stimulate nerve endings and cause pain. Kinins also cause vasodilation and promote chematoxis, that is, a chemical stimulus that promotes the migration of macrophages and other leukocytes to the site of the injury. Swelling, pain, and tissue damage itself lead

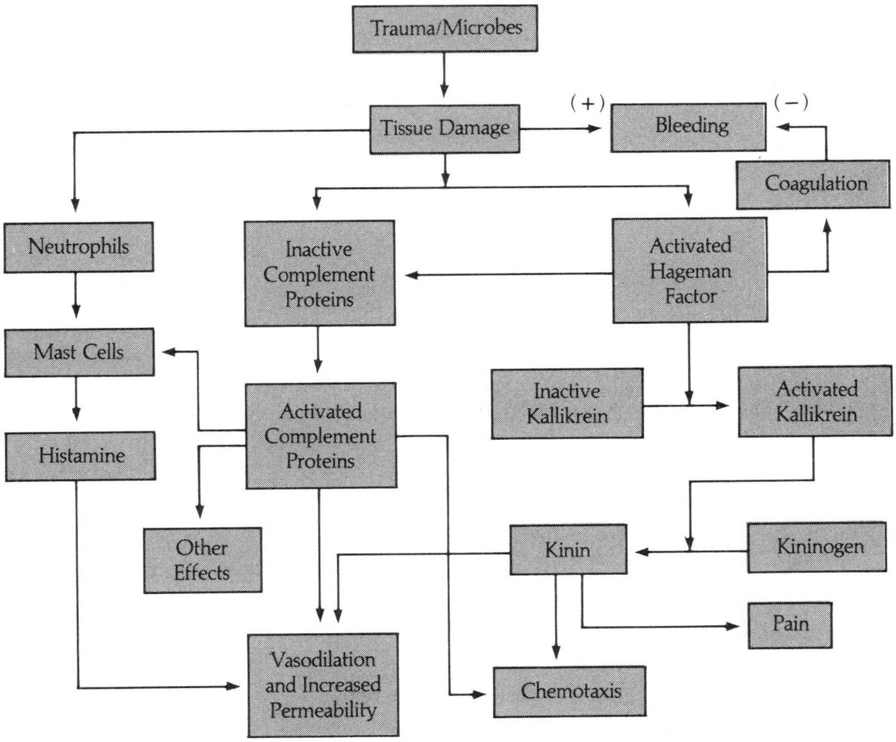

FIGURE 7-6 Pathways involved in the inflammatory response. Details of this response are discussed in the text. Chemotaxis is the process by which leukocytes are drawn to the inflamed site by chemical mediators.

to a variable degree of loss of function in the inflamed region. Swelling, which is localized edema, is brought about by vasodilation of the resistance vessels in the microcirculatory bed and increased permeability of the capillaries to fluid and plasma proteins.

Fluid balance across capillary walls was described in Chapter 2. According to the Starling hypothesis, net flow across the capillary membrane is a function of net filtration pressure and the filtration coefficient, K_f:

$$\dot{Q}_{net} = K_f[(P_C - P_{ISF}) - (\Pi_C - \Pi_{ISF})] \tag{7-1}$$

The terms in brackets are, collectively, the net filtration pressure (see Chapter 2). The permeability of the capillary is part of the filtration coefficient. Normally, there is net flow ($\dot{Q}_{net}$) of fluid out of the capillary into ISF. When the permeability of the capillaries increases, K_f increases and net flow increases. The increased permeability also permits proteins to diffuse in increased quantity from the plasma to the ISF. The proteins increase the oncotic pressure of ISF (Π_{ISF}), thereby increasing net filtration pressure. Hyperemia increases blood volume in the capillaries, and therefore hydrostatic pressure (P_C) also increases. The latter effect increases net filtration pressure. The net result of the increased hydrostatic pressure in capillaries and increased oncotic pressure of proteins in ISF is increased filtration of fluid from capillaries, which causes localized edema. The increased permeability of the capillaries also ensures that complement proteins, Hageman factor, antibodies, and other key proteins are delivered to the site of tissue damage.

The actual volume of fluid lost from the vascular space depends to a large extent on the amount of tissue damaged. Except for atopic reactions (described later), the amount of fluid lost is proportional to the size of the injury. Thus, small injuries such as minor cuts or bee stings produce very localized swelling. If a large mass of tissue is damaged, such as occurs with large burns or when infection spreads in the tissues and produces extensive cellulitis, then larger volumes of fluid will be lost from the vascular space.

Following surgical trauma, there is a period of time when capillary permeability is increased because of the inflammatory response. During this period fluid escapes from the vascular space, producing edema in and around the surgical site. Several liters can be lost into the ISF and TCW, especially with extensive surgical trauma. Fluid shifts of this magnitude can lead to hypovolemic shock. Therefore, following surgery, a fluid such as normal saline or Ringer's lactate solution (Table 7-3) is given to replace vascular volume losses to the ISF in the surgically damaged region. Consequently, the volume of fluid given to a patient after surgery will be greater than output (positive external balance) until the inflammatory response subsides. The extra fluid given in these cases is usually mobilized from the ISF after about 72 hours. The kidneys will excrete this excess fluid, and during this period of time, output will exceed input (negative external balance) until the volume of fluid contained in the ISF is again normal. These events are summarized in Figure 7-7.

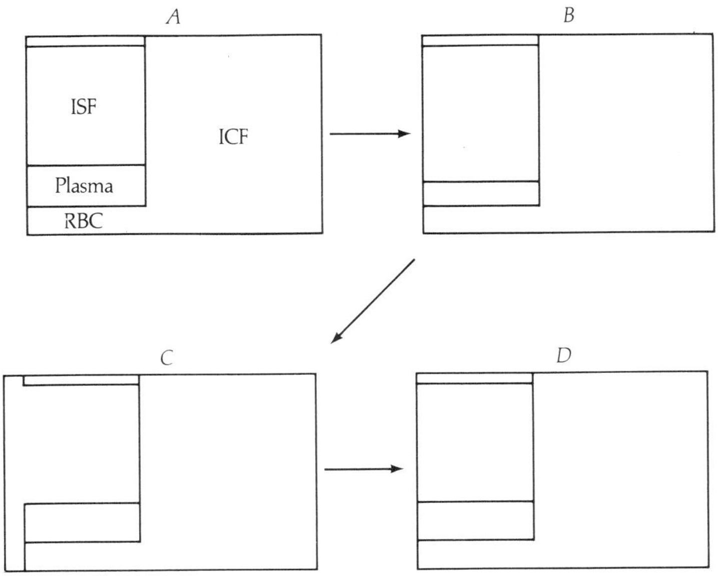

FIGURE 7-7 Fluid shifts following surgical procedures. (*A*) The normal relationships; (*B*) plasma volume has decreased because of the inflammation following surgical trauma (third space loss); (*C*) plasma volume has been restored following infusion of an appropriate infusion fluid. However, because the capillaries still remain more permeable than normal, some of the added fluid expands the ISF even further than that shown in *B*; (*D*) inflammation has resolved and excess fluid has been removed through the kidneys.

A term used in clinical settings—*third space loss* or *third spacing*—indicates loss of fluid from the circulatory or vascular space to other compartments. Third space loss includes processes such as formation of edema, loss of fluid from the vascular space following surgery or burns, formation of ascites, or abnormal accumulation in any compartment. Usually the term is used to imply loss of fluid from the vascular space to another area where the fluid cannot easily exchange with vascular volume.

Occasionally there is an amplification of the local inflammatory response following a minor injury. For example, people who have allergies to bees' venom or other substances may develop a severe and rapidly developing hypotension called anaphylaxis or anaphylactic shock. Evidently, this atopic (uncommon) response develops because of an allergy to the agent causing the injury. Typically, what happens in the generation of anaphylaxis is generalized vasodilation of arterioles throughout the body and constriction of bronchial smooth muscle, causing onset of asthma. The vasodilation, which is caused by release of histamine from mast cells, causes hypotension and loss of consciousness. The atopic response is an exaggerated generalized example of the inflammatory processes that can be fatal unless treated promptly. Parenteral administration of epinephrine can reverse the effects of the atopic response.

Edema from Venous and Lymphatic Obstruction

Local or regional edema can be produced when venous or lymphatic obstruction occurs. Venular constriction or obstruction of veins from tumors or blood clots causes increased hydrostatic pressure in capillaries (P_C in Equation 7-1), which in turn increases net filtration pressure. The resulting increased filtration rate can cause edema of the tissues drained by the obstructed vessel. This is another example of alteration in one of the factors of the Starling hypothesis (Equation 7-1). The effects of changing capillary hydrostatic pressure are shown diagrammatically in Figure 7-8.

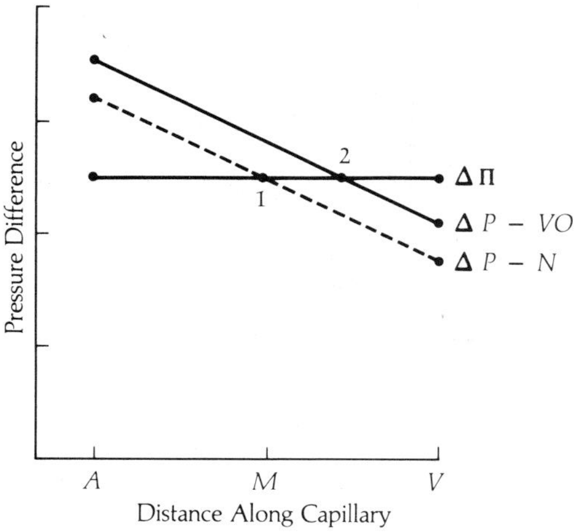

FIGURE 7-8
The effect of increased venular resistance to flow on transmural hydrostatic pressure difference and net filtration pressure in the ideal capillary. When venules constrict or are obstructed (*VO*) the transmural pressure difference (ΔP) increases along the entire length of the capillary. The line representing hydrostatic pressure difference will be shifted above the normal position ($\Delta P - N$) to a new higher position ($\Delta P - VO$) in the diagram. The point at which hydrostatic pressure difference and oncotic pressure difference are equal will be shifted from the midpoint (*M*) of the capillary (point *1*) toward the venular (*V*) end (point *2*). As a result, net filtration pressure will be positive, favoring filtration into the ISF along a greater length of capillary than would normally occur. Compare this figure with the effects of decreased hydrostatic pressure shown in Figure 7-3.

Lymphatic obstruction can lead to profound regional edema. Recall that the lymphatics are the principal pathways by which proteins in the ISF are returned to the plasma. Further, lymphatic drainage returns excess fluid that is filtered from capillaries back into the vascular compartment. Lymphatic vessels are frequently cut and lymph nodes resected during surgical procedures such as radical mastectomy. When lymph nodes are removed in such procedures, the lymphatic drainage from the arm is obstructed, causing brachial edema. The massive swelling seen in elephantiasis is caused by lymphatic obstruction by *Filaria bancrofti*.

Generalized Edema

There are several situations in which edema is not confined to a region but is extensive and spread throughout most of the ISF compartment. Such widespread edema is called *generalized edema* or *anasarca*. The most common pathologies associated with anasarca are the nephrotic syndrome, cirrhosis of the liver, and chronic congestive heart failure. There are specific causes of edema in each of these disease entities.

In the nephrotic syndrome the glomeruli of the kidneys are damaged and there is urinary loss of protein, especially albumin. The protein loss decreases the oncotic pressure of plasma (Π_C, Equation 7-1), and there is increased filtration of fluid from the plasma compartment to the ISF (Figure 7-9). The loss of circulatory plasma volume with its concomitant decrease in blood pressure increases renin secretion by the kidneys and ADH secretion from the neurohypophysis. Hence, sodium and water are retained, as shown by the flow diagrams in Figures 2-6 and 2-7.

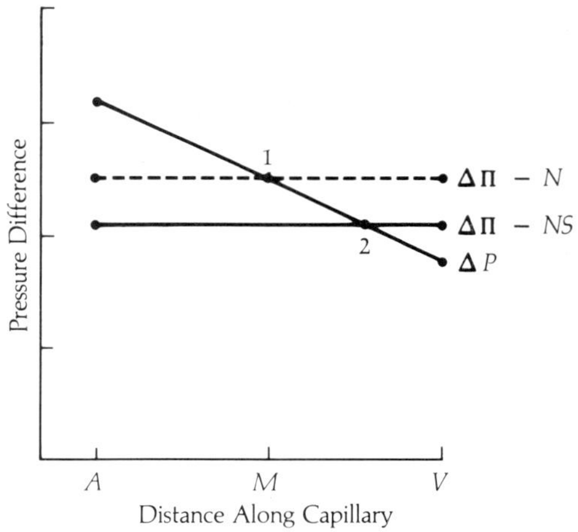

FIGURE 7-9
The effect of decreased colloidal osmotic pressure (oncotic pressure) on net filtration pressure in the ideal capillary. Colloidal osmotic pressure difference ($\Delta\Pi$) is decreased in the nephrotic syndrome (*NS*) because of renal excretion of proteins. Oncotic pressure is also decreased in cirrhosis of the liver. The curve representing normal oncotic pressure difference (dashed line) is shifted down when plasma protein concentration decreases. This shift causes the equilibrium point between oncotic pressure difference and hydrostatic pressure difference to shift toward the venular end (*V*) of the capillary. As a result, filtration of fluid into the ISF proceeds along a greater length of capillary than normal, which favors edema formation.

In cirrhotic liver disease much of hepatocellular function is lost due to the inflammation of the liver and its subsequent scarring. The deficit that is pertinent to this discussion is reduced synthesis of albumin and other serum proteins. In addition, cirrhosis usually obstructs hepatic portal blood flow, causing edema of mesenteric tissues and ascites. Both decreased plasma protein concentration and mesenteric edema lead to reduced blood volume and activation of salt and water retention, as described earlier. Furthermore, it is the liver that is responsible for removing aldosterone from the body. In liver cirrhosis, aldosterone secretion is not only increased, the steroid is not cleared as rapidly as normal from the body. Consequently, there is markedly increased stimulation of the kidneys to retain sodium and water.

In congestive heart failure, perfusion pressure to the kidneys is reduced and renin secretion is increased. The low perfusion pressure is a result of failure of the heart as a pump. Fluid is also lost to the systemic ISF with right heart failure because of increased venous pressure. Venous pressure increases because right ventricular output is reduced and the failing heart requires greater filling pressure to maintain an adequate output. With left heart failure, it is the pulmonary vascular bed that becomes congested and pulmonary edema develops. In either case, following failure, perfusion pressure (blood pressure) to the kidneys decreases and renin secretion by the kidneys increases.

There appears to be a common denominator in generalized edema. In each case, renal perfusion pressure is initially reduced from normal, which causes increased renal secretion of renin. Some authors call this a decrease in the effective circulating blood volume. The key word is *effective*. The control system regulating fluid balance is stimulated because of reduced perfusion pressure in the kidneys, which is initially caused by loss of circulating blood volume to ISF and transcellular fluid (TCW in Figure 1-4). As a result, the renin-angiotensin-aldosterone and ADH controls are stimulated to increase retention of sodium and water. The increased retention creates a positive fluid balance, hence the volume of the entire ECF compartment increases. The steps involved are shown in Figure 7-10.

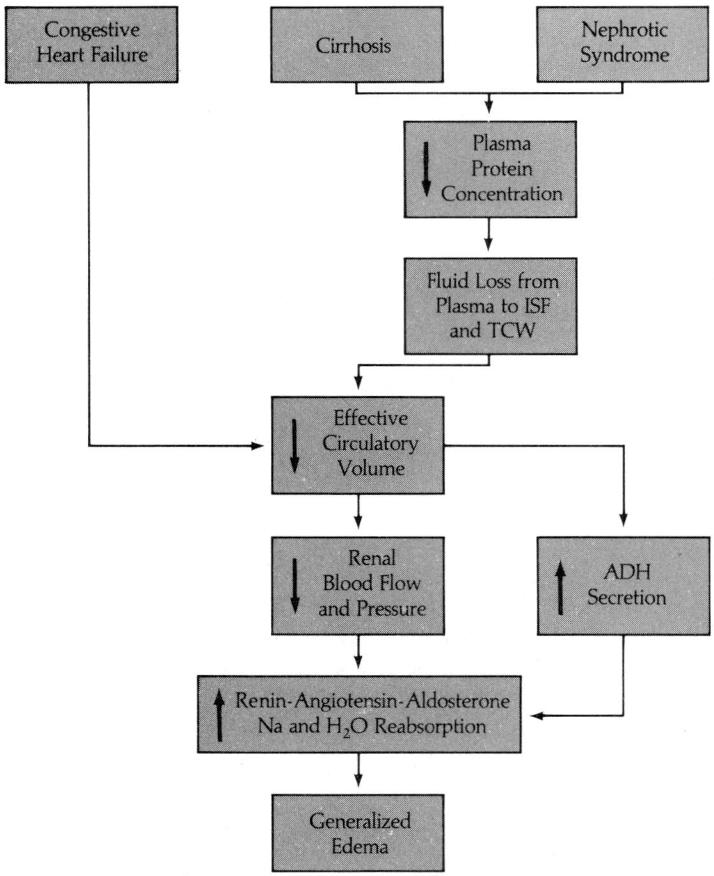

FIGURE 7-10
A flow chart of the steps involved in the production of generalized edema.

Table 7-4 Classification of Disorders of Water Balance

Disturbances of intake	
	1. Water deprivation
	2. Primary polydipsic diabetes insipidus
Disturbances of regulation of output	
	1. Hypothalamic (central) diabetes insipidus
	2. Nephrogenic diabetes insipidus
	3. Syndrome of inappropriate ADH secretion (SIADH)

These fluid balance problems can also lead to derangements of acid-base homeostasis. For example, the stimulus to retain extra sodium causes increased generation of HCO_3^- by the kidneys by means of increased ammonium ion formation and titratable acidity. If the general circulation of blood is not seriously compromised and O_2 delivery to tissues is adequate, the increased HCO_3^- concentration in the ECF will lead to the production of metabolic alkalosis (Chapter 6). On the other hand, if circulation of blood is compromised and there is inadequate delivery of O_2, metabolic acidosis may develop. If left heart failure occurs, producing marked pulmonary edema, a respiratory acidosis may also develop. The type of acid-base abnormality evolving from these conditions that produce generalized edema depends on the severity of the effect on cardiac output and the magnitude of edema produced.

DISORDERS OF OSMOLALITY OF BODY FLUIDS

Regulation of osmolality of body fluids is accomplished by regulating external water balance. The pathophysiology of water balance includes disorders of intake as well as disturbances in the regulation of water output. These disorders may be classified as shown in Table 7-4.

Disturbances of Water Intake

There are two types of derangements of water intake: water deprivation and primary polydipsic diabetes insipidus. Inadequate water intake leads to hyperosmotic volume contraction, as described in the section titled "Volume Contraction." Typically, the individual is either unable to ingest water or does not have fresh water available from the immediate environment.

In the clinical setting water intake may be deliberately reduced either prior to or immediately following surgery to reduce pressure in the vascular system. This is frequently seen in patients who have had either vascular (including open heart) surgery and neurosurgery. Reducing pressure after vascular surgery permits vascular tissues to heal before a high-pressure load is handled. After neurosurgery, blood pressure is reduced and water intake is restricted to minimize the development of cerebral edema.

Excessive water ingestion (polydipsia) from compulsive water drinking may be caused by psychological disturbances, trauma, or tumors in the central nervous system. In these cases, the increased input of water causes inhibition of ADH secretion, which leads to formation of large volumes of dilute urine (diabetes insipidus). The term, *diabetes insipidus*, comes from the Greek word, *diabainein*, which means "to pass through" and from the Latin word, *insipidus*, which means "not savory." Thus, diabetes insipidus is a disease in which large volumes of hypoosmotic (tasteless urine) are excreted from the body. When diabetes insipidus is caused from psychogenic illness, psychiatric treatment is indicated. If the polydipsia is caused by a tumor or another organic disease process, then medical or surgical intervention is indicated.

Disturbances in the Regulation of Output

Antidiuretic hormone is a key regulator of water balance (Chapter 2). If ADH is not secreted at an appropriate rate or if the kidneys do not respond properly to ADH, then output of water from the body cannot be regulated properly. There are three different kinds of disturbances related to regulation of water output. In the first kind an inadequate amount of ADH is available from the neurohypophysis to regulate water excretion. In the second, the kidneys do not respond to the ADH that is present. In the third kind of disturbance, there is an inappropriate secretion of ADH.

Hypothalamic and Nephrogenic Diabetes Insipidus There are two other kinds of diabetes insipidus (besides psychogenic polydipsia) that cause disturbances in the regulation of water excretion. These are hypothalamic and the more rare nephrogenic diabetes insipidus. The hypothalamic type is caused by a deficiency in the secretion of ADH. Without an adequate supply of this hormone, the distal nephron will have a low permeability to water. Under these circumstances, water reabsorption in the distal nephron is reduced and water excretion is thereby increased. In the nephrogenic type, the kidneys do not respond normally to stimulation by ADH. When the kidneys are refractory to ADH stimulation, water will be excreted in excess of normal.

Untreated patients with either type of diabetes insipidus will usually have normal serum sodium concentration and normal extracellular volume. These individuals are able to maintain fluid balance, because the thirst mechanism drives them to ingest large quantities of water to maintain balance. Characteristically, these people are polydipsic because of the polyuria caused by inadequate secretion of or insensitivity to ADH. If patients with these types of diabetes insipidus do not ingest an adequate amount of water, they will develop hyperosmotic volume contraction. Dehydration can occur when these individuals are under enforced restriction of fluids for diagnostic purposes or are incapable of ingesting an adequate volume of water.

Several tests are used that assist in differentiating between the various types of diabetes insipidus. The simplest procedure is to restrict water intake and observe changes in urine flow. If a patient has only primary polydipsic diabetes insipidus, then with water restriction, urine flow will decrease and urine osmolality will increase over a period of 4 or 5 hours to about 600 mOsm/kg water. Normally, maximum urinary concentration is about 1200 mOsm/kg water. However, this concentration will not ordinarily be achieved because previously high urine flow from the polydipsia washes out the concentration gradient in the medulla of the kidney. With partial hypothalamic diabetes insipidus, urine osmolality will increase with fluid restriction but not as much as with primary polydipsia. With complete hypothalamic diabetes insipidus, urine concentration will not change much during the test. Care must be taken to prevent severe dehydration when testing this type of patient with water deprivation. Weight change should be monitored closely and no more than 3 to 5% of body weight should be lost. This is especially important in small children. After water deprivation, patients with hypothalamic diabetes insipidus will respond to subcutaneous or intramuscular injection of ADH by further concentrating their urine, whereas patients with primary polydipsia will not be able to further concentrate their urine after water deprivation. Patients with nephrogenic diabetes insipidus will respond only minimally to exogenous ADH administration.

Syndrome of Inappropriate ADH Secretion (SIADH) An excess of ADH can be present in the serum for a short period of time and produce no apparent clinical problems. However, when the excess ADH remains for about three days, the following six features appear: (1) hyponatremia, (2) the kidneys excrete sodium even in the presence of hyponatremia (sodium wasting), (3) normal tissue turgor and blood pressure, (4) normal BUN and plasma creatinine concentration, (5) inappropriately high osmolality of the urine for the degree of osmolality of body fluids, (6) normal adrenal function.

The syndrome of inappropriate ADH secretion occurs when patients secrete more ADH than is appropriate for the osmolality of their body fluids. Typically these patients have all six features or characteristics of an excess of ADH that are listed above. They have a mild hypoosmotic volume expansion. The increased volume is seldom more than 4 or 5 L total, hence, blood pressure and tissue turgor are normal. These patients will usually have normal volume and composition of urine, but neither the volume nor osmolality of the urine is matched to body fluid conditions. They will usually have normal amounts of sodium in the urine. In addition, if they are given sodium orally or parenterally, they will excrete the added sodium in their urine. In other words, they regulate body osmolality and sodium concentration at new values that are not typical for humans.

Table 7-5 Conditions Associated with SIADH

1. Lung tumors (oat cell carcinoma)	6. Subarachnoid hemorrhage
2. Pneumonia	7. Side effects of certain drugs including:
3. Head injuries	a. chlorpropamide
4. Encephalitis	b. vincristine
5. Brain abscess	c. cyclophosphamide

The hyponatremia is caused by dilution of ECF with water that is retained because of the excess ADH. The sodium wasting is not due to a direct effect of ADH on the renal tubules but rather to dilution of plasma proteins. Recall that osmotic uptake of the sodium chloride solution reabsorbed into the capillary from the ISF surrounding the proximal tubule is a function of the protein concentration of the plasma (Starling hypothesis). The excess water being retained because of the surplus ADH dilutes these proteins, hence less sodium chloride and water will be reabsorbed from the proximal tubule. When extra sodium is given to these patients, the extra sodium and fluid are excreted. The kidneys are responding as if the ECF were volume expanded by excreting excess sodium and water. However, the kidneys are also responding to the excess ADH; hence water is still retained, thereby maintaining the hyponatremia.

There are several causes of SIADH. Head injuries and brain surgery can initiate increased ADH secretion. In addition, pneumonia and ADH-secreting tumors, especially from oat cell carcinoma of the lung, are known to cause SIADH. Other causes are listed in Table 7-5. Treatment of SIADH is usually by water restriction until concentration of body fluids restores osmolality to normal. If the hyponatremia is severe and the patient has signs of central nervous system dysfunction that are associated with hyponatremia, hypertonic saline should be administered along with a diuretic to prevent excess volume expansion of the extracellular compartment. This combination will raise the sodium concentration of ECF and at the same time decrease ECF volume.

Electrolyte Balance and Imbalance

CHAPTER 8

As discussed in previous chapters, diseases of the kidneys, lungs, cardiovascular, and gastrointestinal systems can cause disturbances of fluid, electrolyte, and acid-base regulation. It should also be noted that both concentrations and stores of electrolytes are altered in the presence of those regulatory disorders. Stores of an electrolyte refer to the total amount of that electrolyte present in all forms within the body. Some of the stores are exchangeable through external and internal exchange processes (Chapter 2). Other portions of the store of a given electrolyte may not be exchangeable. The focus of this chapter is on the normal balance and the derangements of electrolyte balance commonly encountered in the clinical setting.

SODIUM

Sodium balance is integrated with regulation of extracellular fluid volume. The various processes that regulate Na^+ concentration are described in Chapters 2 and 7. However, it is useful to describe pathologic situations in which values for Na^+ concentration may be abnormal, because this electrolyte concentration is so easily and frequently measured.

Sodium Stores An adult contains about 60 mEq of Na^+ per kilogram of body weight. Thus, an 80-kg man would contain approximately 4800 mEq of Na^+. About 40% of this Na^+ is found in bone, 50% in the extracellular fluid,

Table 8-1 Causes of Hyponatremia

I. Decreased ECF volume	II. Normal ECF volume
A. Renal losses	A. SIADH
1. Diuretics	B. Glucocorticoid deficiency
2. Adrenal insufficiency	C. "Reset osmoreceptors"
3. Renal disease	III. Increased ECF volume (generalized edema)
B. Nonrenal losses	A. Cirrhosis
1. Gastrointestinal	B. CHF
2. Third space sequestration	C. Renal disease
3. Skin	

and the remaining 10% in the intracellular fluid. Nearly 80% of the total Na^+ stores is exchangeable. The exchangeable fraction includes all of the intracellular and extracellular Na^+ plus about half of the Na^+ in bone. The nearly 20% that is not exchangeable is adsorbed on the surface of hydroxyapatite crystals in the matrix of compact bone. The exchangeable portion is available for replacement of *some* of the Na^+ that is lost in sweat, urine, diarrhea, or gastric drainage. In addition, when Na^+ is retained with volume expansion it is distributed with the exchangeable fraction of Na^+ stores.

Hyponatremia

Hyponatremia is said to be present when Na^+ concentration of extracellular fluid is less than 136 mEq/L. This condition can occur with a reduced, normal, or expanded extracellular fluid volume, as shown in Table 8-1. There are also situations in which analysis of serum indicates hyponatremia but Na^+ concentration of ISF is normal. Each of these situations is discussed in the following sections.

Hyponatremia with Volume Contraction There are two general categories of hyponatremia with contraction of extracellular fluid volume, classified as renal losses and nonrenal losses (Table 8-1). Excessive loss of Na^+ through the kidneys can occur with use of diuretics, with adrenal insufficiency (Addison's disease), and with renal disease itself. Nonrenal losses occur from the gastrointestinal tract, through the skin, and with third space sequestration.

When hyponatremia and volume contraction are caused by renal loss, the kidney must excrete Na^+, its accompanying anions, and water into the urine and yet retain enough water to produce hypoosmotic volume contraction (see Figure 7-2*C*). Normally, the kidneys rid the body of excess water by reabsorbing solute and leaving the excess water behind in the tubular fluid. This activity occurs in the diluting segments of the nephron, namely, the loop of Henle, the distal tubule, and the collecting duct. The process is called generation of "free water," which means the kidneys are producing water for excretion. Ordinarily, the free water is excreted in the urine.

Diuretic agents can interfere with generation of free water. These drugs are used clinically to decrease the volume of the extracellular fluid. They act by reducing or blocking Na^+ (or chloride) reabsorption in the diluting segments of the nephron. With initial use of diuretics, more Na^+ and its anions remain in the tubular fluid than normal. This, in turn, reduces water reabsorption and causes diuresis.

Prolonged use of diuretics can lead to hyponatremia for two reasons. First, extra Na^+ and water are lost in urine, producing volume contraction. The reduced Na^+ reabsorption itself tends to promote hyponatremia. However, a response to volume contraction is increased release of ADH, which in turn promotes water retention out of proportion to the need for maintaining normal osmolality of ISF. Second, most diuretics cause hypokalemia, which, when severe, increases the sensitivity of osmoreceptors in the hypothalamus to the osmolality of ISF in the brain. Hypokalemia, thus, also promotes more ADH release than is needed for the prevailing osmolality of body fluids. Therefore, prolonged use of diuretics leads to loss of Na^+ and its anions into the urine, loss of some ECF volume, and then retention of water at lower ECF volumes. These effects, taken together, cause hyponatremia with volume contraction.

When there is a deficiency of aldosterone, Na^+ and its anions are lost in the urine. The volume contraction produced stimulates ADH secretion, and the result is essentially the same as that just described. In renal disease, damage to the tubules causes malfunction of Na^+ transport and increased Na^+ excretion. The disease usually disrupts function in the loop of Henle and there is a diminished ability to concentrate or dilute urine. In renal disease, urine tends to have higher concentrations of Na^+ than normal and has an osmolality close to or equal to that of plasma (isosthenuria).

Hyponatremia is also associated with nonrenal causes of Na^+ loss. Sodium is frequently lost with gastrointestinal secretions, for example, with diarrhea and gastric suction. Sodium will also be lost with fluid when the integumentary seal is damaged, as occurs in burns or trauma. If losses are sufficient to reduce ECF volume enough to impair renal blood flow, then the hypovolemia plus impaired renal excretion of water can lead to hyponatremic volume contraction. Antidiuretic hormone plays the same role in nonrenal losses of Na^+ as it does with renal losses.

Plasma fluid volume can decrease with third space sequestration of fluid such as occurs with pancreatitis and peritonitis. In these examples, when the volume of fluid contained in the third space increases, fluid shifts from plasma to the transcellular compartment. There is actual loss of plasma volume, but total body fluid volume may not change because the lost volume from plasma is only shifted across a membrane to another compartment. The hypovolemia again causes ADH release as described earlier. In this latter case, however, total body fluid volume may not have changed.

Hyponatremia with Normal ECF Volume The most common cause of hyponatremia in the presence of a normal ECF volume is the syndrome of inappropriate ADH secretion. The mechanisms and characteristics of SIADH were discussed in Chapter 7. The only point that should be reiterated here is that with SIADH the ECF volume is essentially normal or at most only mildly increased (4 or 5 L maximum). For this reason, SIADH is also classified as hyponatremia with normal ECF volume.

When there is insufficient secretion of glucocorticoid hormones, patients develop hyponatremia and a pattern of characteristics that mimics SIADH. The main difference is that in these individuals, adrenal function is abnormal and replacement of glucocorticoids results in production of dilute urine and correction of hyponatremia. Evidently, decreased glucocorticoid secretion causes abnormal secretion of ADH, which is corrected with glucocorticoid replacement.

Some patients with chronic and debilitating disease develop hyponatremia. In these situations, the hyponatremia is seldom severe and patients can concentrate and dilute their urine normally. However, the osmoreceptors in the hypothalamus secrete ADH at lower osmolalities than normal. It is as though these receptors were reset to regulate osmolality of body fluids at a lower than normal setpoint, for example, 270 instead of 290 mOsm/kg of water. These people seldom have to be treated for hyponatremia per se.

Hyponatremia with Increased ECF Volume Patients with chronic congestive heart failure, cirrhosis, or nephrotic syndrome will sometimes develop hyponatremia in association with their expanded ECF volume. Usually the hyponatremia occurs in the late stages of these diseases and is associated with a poor prognosis. The mechanism of volume expansion in these patients with generalized edema was discussed in Chapter 7. The cause of hyponatremia in these patients may be reduced renal perfusion and excess ADH, but no clear evidence has been published to support this hypothesis.

Psychogenic polydipsia is also associated with an expanded ECF volume and hyponatremia. However, water restriction generally corrects both volume and osmolality abnormalities in these cases.

Patients with chronic renal failure may drink only moderate amounts of water and develop hyponatremia with expansion of ECF volume because their kidneys cannot excrete water loads rapidly. Again, if symptoms are not too severe, water restriction will correct the problem.

Hyponatremia from Redistribution of Water When nonsodium solutes accumulate in the ECF, water will shift from the ICF to the ECF by osmosis and osmotic equilibrium will be maintained (Figure 8-1). This situation occurs clinically with diabetes mellitus when the glucose concentration of ECF increases markedly beyond normal ranges. The extra glucose causes osmosis of water from the ICF compartment, which in turn dilutes the Na^+ in the ECF. The process itself leads to hyponatremia (see Appendix 5). The initial shift of water from the intracellular compartment will increase the volume of the ECF

FIGURE 8-1
Hyponatremia from redistribution of water. (*A*) Glucose concentration of ECF (*G*) has increased above normal. The increased osmolality causes osmosis from the ICF until osmolality of ICF and ECF are equal. The shift of water from the ICF dilutes the sodium in the ECF causing hyponatremia. (*B*) The increased glucose concentration has persisted causing osmotic diuresis, which reduces ECF volume. The effect on sodium concentration will depend on the relative loss of sodium to volume in the urine and the amount of sodium intake during the period of diuresis. Heavy lines indicate the normal volume and osmolality of body fluids.

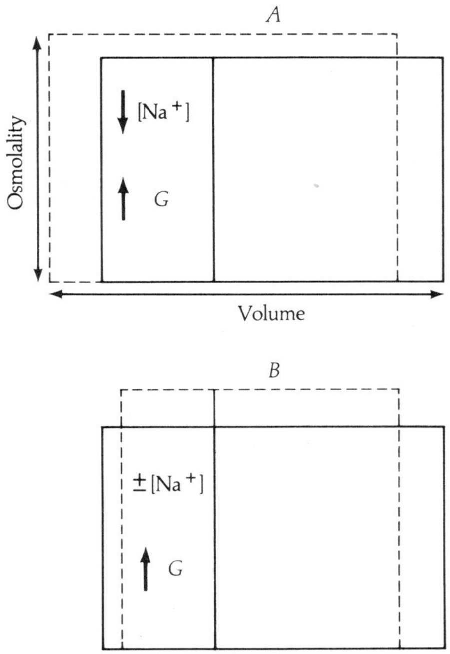

(Figure 8-1*A*). However, if the increase in glucose concentration in the ECF is allowed to remain, the glucose creates an osmotic diuresis that contracts extracellular volume and also causes a loss of Na^+ in the urine. The loss of water will exceed the loss of Na^+. The net results will be hyperosmotic volume contraction (Figure 8-1*B*). The final Na^+ concentration may be normal, less than normal, or greater than normal depending on volume changes relative to Na^+ loss. In this situation, correction of the increased glucose concentration with insulin must be accompanied by both Na^+ and volume replacement. An example of this type of fluid and electrolyte abnormality is presented in Chapter 10.

Pseudohyponatremia The Na^+ concentration of ECF is estimated from the concentration of Na^+ in plasma or serum. The concentration is usually expressed per liter of plasma or serum. It should be noted that the plasma concentration of Na^+ is less than that of ISF by approximately 5 mEq/L (see Table 1-3). This difference is caused by two factors. First, ions in ISF are in a specific kind of equilibrium with ions in plasma water called Donnan equilibrium. With Donnan equilibrium cation concentration will not be the same on both sides of the capillary membrane. Donnan equilibrium is tangential to the present discussion, but is presented in Appendix 4. Second, the bulk of protein in plasma dilutes the plasma water. The Na^+ (and other ions) are dissolved only in the water phase and not in the protein phase. However,

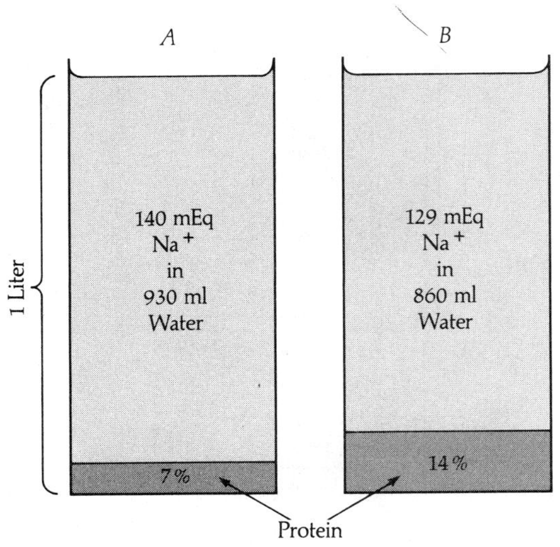

FIGURE 8-2
Pseudohyponatremia. (*A*) Sodium and protein concentration are normal; (*B*) protein concentration is doubled, thereby decreasing plasma water volume. See the text for a detailed explanation.

concentration of Na^+ is expressed per liter of plasma as though the Na^+ were distributed evenly throughout both the water and protein phases (Figure 8-2*A*). In fact the 140 mEq of Na^+ in each liter of plasma is dissolved in the water phase alone. Because water is approximately 93% of plasma volume, the 140 mEq is dissolved in only 0.93 L of water. Thus, in the example shown in Figure 8-2*A*, the concentration of Na^+ in plasma water is

$$\frac{140 \text{ mEq}}{0.93 \text{ L}} = 150 \text{ mEq / L plasma water} \qquad (8\text{-}1)$$

There are diseases in which the protein or lipid concentration of plasma increases markedly above normal values. An example of the effects of increased protein concentration is shown in Figure 8-2*B*. In this example, protein concentration is doubled from 7% protein to 14%. If the increased protein concentration is the only abnormality, the Na^+ concentration would be 129 mEq/L of plasma. This appears to be hyponatremia, but, in fact, the concentration in plasma water is normal. The 129 mEq of Na^+ is distributed only in the water phase, which accounts for 86% of the total volume of plasma. The concentration of Na^+ per liter of plasma water is

$$\frac{129 \text{ mEq}}{0.86 \text{ L}} = 150 \text{ mEq / L plasma water} \qquad (8\text{-}2)$$

Thus, the concentration of Na^+ in plasma water is normal. Because Na^+ in the ISF is in equilibrium with Na^+ in the water phase of plasma, the concentration of Na^+ in the ISF will be normal and no signs or symptoms of hyponatremia will be present. This situation is called pseudohyponatremia or factitious hyponatremia, because Na^+ concentration in the extracellular water is normal.

Pseudohyponatremia occurs when the concentration of lipids or proteins in the serum is increased above normal. Hyperlipemia can cause pseudohyponatremia in diseases such as diabetes mellitus, nephrotic syndrome, and some types of cirrhosis of the liver. The serum obtained from patients with hyperlipemia is often markedly lactescent. Hyperproteinemia of sufficient magnitude to produce pseudohyponatremia occurs in multiple myeloma and is relatively rare.

Hypernatremia

Hypernatremia (serum Na^+ concentration greater than 145 mEq/L) occurs when hypoosmotic fluid is lost from the body. This type of loss is classified as hyperosmotic volume contraction (Chapter 7). Usually, loss of hypoosmotic fluid from sweating or vomiting produces a mild hypernatremia that is ordinarily corrected with fluid ingestion. Severe hypernatremia is present when the Na^+ concentration in plasma is 160 mEq/L or greater (greater than 172 mEq/L of plasma water). Increased Na^+ concentrations to these extremes occur when there is complete absence of fluid intake, as might occur when patients cannot ingest fluid because of a stroke, surgery, or coma. It also occurs with inadequate replacement of water in diabetes insipidus and with osmotic diuresis. The increased osmolality of ECF leads to water shifts from the ICF by osmosis and, therefore, causes cellular dehydration as well (see Figure 7-2*D*).

Osmotic diuresis can be produced in patients who receive tube feedings that contain high concentration of protein. The urea produced from metabolism of the protein can cause an osmotic diuresis that can lead to production of a severe hypernatremia.

It should be noted that physical signs and symptoms of hypernatremia are not clearly defined nor are they specifically diagnostic. Patients with this electrolyte disorder will be seriously ill from other disease processes. The diagnosis of hypernatremia must be made or at least confirmed from measurement of serum Na^+ concentration. Treatment consists of replacing the water that has been lost either orally or through intravenous administration of 5% glucose solution.

Another cause of hypernatremia is Na^+ overload. This can occur clinically with acute ingestion or infusion of hypertonic solutions of sodium chloride or sodium bicarbonate. Frequently, hypernatremia is produced following infusion of Na^+ bicarbonate to counteract metabolic acidosis during cardiac arrest. This is a special problem for infants in whom as little as 100 mEq of Na^+ can increase plasma Na^+ concentrations to 160 mEq/L. Causes of hypernatremia are summarized in Table 8-2.

Table 8-2 Causes of Hypernatremia

I. Hypoosmotic fluid loss (hyperosmotic volume contraction)	II. Increased sodium intake
A. Increased insensible loss	A. Hypertonic sodium chloride solutions
1. Burns (damage to integumentary seal)	B. Sodium bicarbonate
2. Fevers	
3. High temperature environments	
4. Exercise	
B. Renal loss	
C. Diabetes insipidus	

POTASSIUM

The highest concentrations of K^+ are found in the ICF. The average concentration in ICF is probably close to 150 mEq/L. This cation plays a major role in regulating the excitability of nerve and muscle cell membranes, activating specific enzymes in the intracellular environment, and in the normal functioning of the kidneys. The usual dietary intake of K^+ varies from about 40 to 120 mEq/day. To maintain K^+ balance and a constant concentration in the ECF, the excretion of this cation must equal the intake. There are two steps in the balance process that tend to reduce K^+ concentration in ECF, namely, cell uptake and excretion. After absorption from the alimentary canal, K^+ enters cells (Figure 8-3). The entry of K^+ into skeletal muscle and liver cells is partially insulin dependent. Furthermore, it has been shown that insulin secretion is increased by increased K^+ concentration in plasma and vice versa. The importance of cell uptake may be illustrated by what happens to the K^+ in food eaten at mealtime. If 40 mEq of K^+ from a meal were added to the ECF without cell uptake, the extracellular concentration of K^+ would increase from 4 mEq/L to as high as 7 mEq/L, depending on body size. Therefore, uptake of ingested K^+ by cells minimizes the increase in K^+ concentration of ECF after a meal.

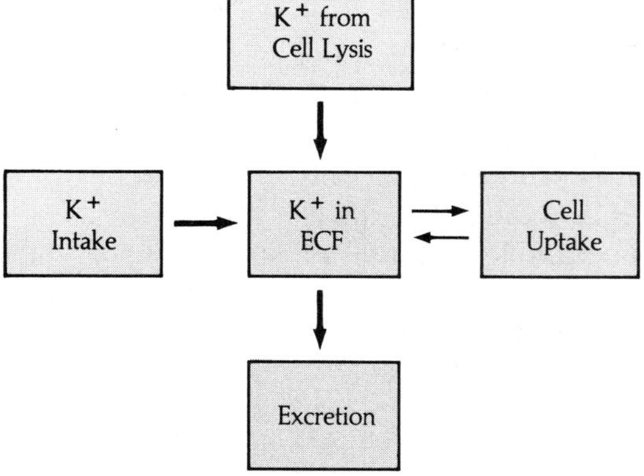

FIGURE 8-3
Processes involved in potassium balance. Cell uptake and loss is a major step in regulating potassium concentration in the ECF. When potassium intake increases, for example, during absorption of a meal, cells take up potassium and then release it slowly as the kidneys excrete the excess. Potassium may be lost from the ICF when potassium excretion is excessive, for example, during metabolic alkalosis and acidosis. Thus, serum concentration can be maintained within normal limits even when total body stores are decreased or increased.

Renal excretion is slower than cell uptake, but it is the process by which balance with intake is maintained. Renal handling of K^+ is a precisely regulated function that involves aldosterone and the Na^+ concentration in the fluid of the distal nephron. Because nearly all of the K^+ filtered at the glomerulus is reabsorbed in the proximal tubule, the K^+ appearing in the urine must be secreted by cells in the distal segments of the nephron. Fecal excretion of K^+ normally accounts for only a small percentage of the total K^+ lost from the body per day and ordinarily plays only a minor role in K^+ balance.

Potassium Stores The adult body has, on the average, about 50 mEq of K^+ per kilogram of body weight. Therefore, an 80-kg male body would contain 4000 mEq of K^+. Nearly all of this amount is located in the intracellular fluid. Of the total, only 65 mEq (about 1.6%) is dissolved in the extracellular fluid. Virtually all of the K^+ in the body is exchangeable. The rather precise regulation of K^+ concentration in the ECF and the exchangeability of K^+ can lead to difficulty in assessing K^+ balance. For example, the increased hydrogen ion concentration of acidosis leads to $K^+ - H^+$ exchange across cell membranes and loss of K^+ from cells (see Figures 5-4 and 5-9). However, the kidneys will excrete the extra K^+ shifted from ICF and plasma K^+ concentration does not increase. Only after prolonged acidosis or correction of acidosis will the deficit in total stores be manifested as a decrement in serum concentration of K^+. In alkalosis, K^+ is also lost from intracellular stores, yet until these losses are fairly large, serum K^+ concentration will be normal. Therefore, unless changes in K^+ stores are of sufficient magnitude, these changes will not necessarily be reflected by alterations in the concentrations of K^+ in serum. On the other hand, when changes are of sufficient magnitude, hyperkalemia or hypokalemia will occur, and both can be life threatening.

Hypokalemia

Hypokalemia is present when plasma K^+ concentration is less than 3.5 mEq/L. There are three principal causes of hypokalemia. These are: (1) increased cellular uptake, (2) excess renal excretion, and (3) excessive gastrointestinal loss (Table 8-3).

Increased Cellular Uptake of Potassium Cells increase their uptake of K^+ when there is excess insulin, in alkalosis, and in certain specific diseases. Hypersecretion of insulin can occur with high carbohydrate intake, especially with intravenous hyperalimentation. Solutions containing high concentrations of glucose are also frequently infused with added insulin. Both insulin and glucose promote K^+ uptake by cells, and this can lead to a transient hypokalemia.

In alkalosis, hydrogen ions from the ICF exchange across cell membranes for Na^+ and K^+ in the ISF (see Figures 5-4 and 5-9). This exchange reduces K^+ concentration of ECF (hypokalemia). If the alkalosis persists, the kidneys

Table 8-3 Causes of Hypokalemia

I. Increased cellular uptake	III. Excessive gastrointestinal losses
A. Excess insulin	A. Gastric loss
B. Alkalosis	1. Vomiting
II. Excessive renal excretion	2. Nasogastric suction
A. Hyperaldosteronism	B. Diarrhea
1. Volume depletion	
2. Mineralocorticoid-secreting tumors	
B. Increased sodium delivery to distal nephrons	
1. Diuretics	
2. Nonabsorbable ions	
C. Acid-base disturbances	
1. Metabolic acidosis and alkalosis	
2. Respiratory acidosis and alkalosis	

excrete excess K^+, which can cause depletion of K^+ stores. When K^+ depletion and hypokalemia are severe in the presence of alkalosis, the kidneys will attempt to conserve K^+ by exchanging renal intracellular hydrogen ions for Na^+ in tubular fluid (see Figure 4-9). The net result is generation of extra bicarbonate, which aggravates the alkalosis and produces an acid urine even though ECF is alkalotic. This situation is called paradoxical aciduria and is usually associated with marked K^+ depletion.

Excess Renal Excretion Three conditions cause increased renal excretion of K^+: (1) hyperaldosteronism, (2) increased Na^+ and fluid delivery to the distal nephron, and (3) acid-base disturbances. Aldosterone stimulates Na^+ reabsorption and K^+ secretion in distal nephron cells. Because volume depletion is a powerful stimulator of aldosterone secretion (through renin-angiotensin), volume depletion itself can lead to K^+ loss and hypokalemia. In addition, secretion of excess mineralocorticoid accentuates the process leading to Na^+ retention and K^+ loss. For example, some of the intermediates in the synthetic pathway for aldosterone are biologically active. Tumors of the adrenal cortex may secrete several steroid substances that possess mineralocorticoid activity. It is known that the steroid, deoxycorticosterone, (see Figure 2-8) also stimulates Na^+–H^+ exchange by distal nephron cells and can cause metabolic alkalosis when given as a drug. Aldosterone by itself does not usually produce as severe an alkalosis as deoxycorticosterone given in the same dose. Both steroids cause K^+ depletion and alkalosis when they are present in the blood in higher than normal concentrations.

As described in Chapter 4, increased delivery of sodium to distal nephron sites causes the tubular lumen to become more electronegative (see Figure 4-12). The increased delivery of Na^+ is also accompanied by increased fluid flow in the distal nephron. The increased volume of fluid tends to dilute the secreted K^+ and thus helps maintain the concentration gradient for K^+ between the renal tubular cells and tubular fluid. Therefore, increased fluid flow and electronegativity can promote K^+ secretion into the lumen of the distal nephron. Diuretics are the drugs most commonly used to increase sodium and fluid delivery to the distal nephron. Poorly reabsorbed anions (accompanied by Na^+) such as $HPO_4^=$ or the conjugate bases of ketoacids, or antibiotics that are anions (such as carbenicillin) also cause increased Na^+ and fluid delivery to the distal nephron. In each case, the resulting increased K^+ secretion can cause depletion of K^+ stores and hypokalemia.

Acid-base disturbances also cause K^+ depletion. In metabolic acidosis, conjugate bases of the fixed acids constitute poorly reabsorbed anions, which cause increased delivery of Na^+ to the distal nephron. In addition, in persistent respiratory and metabolic acidosis, K^+ exchanges for H^+ across cell membranes and the K^+ extruded into the ECF is excreted. In respiratory acidosis, the K^+ depletion may not be apparent until the acidosis is treated. Bicarbonate given to increase pH will reverse the H^+–K^+ exchange process and cause hypokalemia. In metabolic alkalosis, the increased filtration of bicarbonate causes increased delivery of fluid to the distal nephron. Again, this process promotes K^+ secretion and loss. Profound K^+ depletion can occur with metabolic alkalosis.

Gastrointestinal Losses The K^+ concentration of saliva, gastric juice, and ileal and colonic fluid is greater than in plasma. Hence, fluid loss from the alimentary canal frequently leads to substantial K^+ loss and hypokalemia. When significant quantities of gastric juice are removed by nasogastric suction, the K^+, Cl, and Na^+ should be replaced in a volume of liquid equal to that removed. This regimen generally prevents the development of severe metabolic alkalosis and K^+ depletion. Diarrhea, especially in children, can cause volume depletion and metabolic acidosis from bicarbonate losses as well as hypokalemia. With prolonged bouts of diarrhea, replacement of both volume and electrolytes is often necessary to prevent hypovolemic shock and hypokalemia.

Hyperkalemia

Hyperkalemia can be generated by four processes: (1) increased intake of K^+, (2) increased cell lysis, (3) altered cellular uptake, and (4) decreased renal excretion (Figure 8-3 and Table 8-4).

Increased Intake Increased dietary intake of K^+ is a rare cause of hyperkalemia and ordinarily cannot occur without concomitant renal failure or decreased cellular uptake such as could occur with a lack of insulin. Intravenous infusion of K^+ represents another intake process that can cause hyperkalemia. A bolus of a drug that contains K^+, such as penicillin, or infusion of extra K^+ in parenteral solutions can cause a transient hyperkalemia. Even transient hyperkalemia can be dangerous because the condition can cause cardiac arrest; hence, care must be taken when infusing solutions containing K^+.

Red cells in stored blood tend to release K^+ into the plasma. After three weeks of storage the K^+ concentration in the plasma can be as much as 30

Table 8-4 Causes of Hyperkalemia

I. Increased intake of K^+	IV. Decreased renal function
A. Oral (only in renal failure or lack of insulin)	A. Renal failure
B. Intravenous infusion	B. Volume depletion
II. Increased cell lysis	V. Pseudohyperkalemia
III. Altered cellular uptake	
A. Tissue catabolism	
B. Acidosis	
C. Lack of insulin	

mEq/L. However, unless the blood is infused very rapidly in people with normal kidney function, the recipient would not experience a significant increase in K^+ concentration in the plasma. On the other hand, if the patient were in renal failure, the added load of K^+ could cause a fatal hyperkalemia.

Increased Cell Lysis Increased breakdown of tissues (tissue catabolism) with the release of intracellular K^+ can cause hyperkalemia, especially if there is concomitant renal failure. A common example of this cause of hyperkalemia is in trauma from motor vehicle accidents, and crush syndrome from other kinds of trauma. Surgical procedures may cause hyperkalemia because of the leakage of K^+ from damaged tissues. This process is particularly important for patients with renal failure. Patients with lymphomas can develop hyperkalemia after treatment with cytotoxic drugs, especially when the tumor mass is large.

Altered Cellular Uptake Acidosis frequently causes an increase in the K^+ concentrations of ECF because the H^+–K^+ exchange across cell membranes causes loss of K^+ from the ICF as H^+ concentration increases. Again, if renal function is normal, the serum K^+ concentration may not be increased significantly. However, if the acidosis is caused by renal failure, then the acidemia may cause marked hyperkalemia.

Lack of adequate pancreatic secretion of insulin will slow much of the uptake of K^+, which normally occurs with absorption of K^+ from the alimentary canal or when K^+ is added from other routes. Thus, patients with diabetes mellitus are more likely to develop hyperkalemia than people with normal insulin secretion and concentration in plasma. This effect is especially pronounced in diabetics who develop ketoacidosis. Renal excretion may be the only factor that prevents a diabetic in ketoacidosis from developing fatal hyperkalemia.

Decreased Renal Excretion Renal failure is a major cause of hyperkalemia. The kidneys are the only significant route by which K^+ can be eliminated from the body. Therefore, chronic or acute loss of adequate renal function will cause K^+ retention. In chronic failure, K^+ concentration does not increase until oliguria develops. The mechanism by which the kidneys are able to adapt in chronic nonoliguric renal failure is not known. As long as the diet remains normal the functioning renal tissue will maintain K^+ balance. However, if the load of K^+ is increased suddenly, the diseased kidneys require a longer period of time to excrete that load. There may, as a result, be a period of hyperkalemia until the excess K^+ is excreted. Once chronic oliguric renal failure sets in, K^+ balance becomes increasingly difficult to maintain and hyperkalemia becomes more difficult to prevent. Finally, only dialysis will permit maintenance of K^+ balance. In acute renal failure the kidneys cannot maintain normal K^+ balance, and hyperkalemia can develop very quickly. Potassium restriction and frequently dialysis are required to prevent hyperkalemia with acute renal failure until the kidneys have recovered sufficiently to again maintain balance.

Both volume contraction and reduction in effective circulating volume can decrease K^+ excretion by reducing renal perfusion. When severe, decreased blood flow causes prerenal failure, which in turn causes hyperkalemia. Correction of the fluid deficit or improvement in renal blood flow will usually improve K^+ excretion and permit regulation of K^+ balance. However, if the renal blood flow is decreased for too long a period of time, acute renal failure (acute tubular necrosis or ATN) may result.

Pseudohyperkalemia When K^+ concentration is measured clinically, it is measured in serum and ordinarily not as plasma K^+ concentration. To obtain serum, a sample of blood is allowed to clot and when the clot retracts, serum is expressed and may be sampled for analysis. Normally, K^+ is released from white blood cells (WBCs) and platelets during clotting and this release may increase serum concentrations by as much as 0.5 mEq/L. However, in patients with high white blood counts (leukocytosis greater than 100,000 WBCs/μL) or very high platelet counts (thrombocytosis greater than 500,000 platelets/μL) the K^+ released with clotting may cause serum concentrations to increase to as high as 7 to 9 mEq/L. The patient will exhibit no other signs or symptoms of hyperkalemia. If a sample of blood from such patients is drawn into a tube with an anticoagulant such as heparin and the plasma is analyzed, the K^+ concentration will be found to be normal. In addition, if red blood cells lyse and release their contents into the serum before analysis, the K^+ released from them will also cause a false increase in serum K^+ concentration. In this latter case, however, the serum will have a red tinge as a clue to the cause of the apparent hyperkalemia.

Physiologic Effects of Disturbed Potassium Balance

Both hyperkalemia and hypokalemia can cause dysfunction of excitable membranes of muscle and neural tissue. The kidneys are also adversely affected by changes in K^+ concentration. In order to understand the effects of disordered

K^+ balance on excitable tissues, it is first necessary to review electrical characteristics of excitable cells. All cells are electrically charged *with the inside of the cell negative* with respect to the outside of the cell. For example, nerve cells have a voltage (or potential difference) of 70 millivolts (mV) across their cell membranes whereas skeletal and cardiac muscle cells have a potential difference close to 90 mV. As long as the muscle and nerve cells are not active, these potential differences will be stable and are called resting potentials. The symbol for resting potential is E_m. The magnitude of the resting potential is proportional to the ratio of the K^+ concentration inside the cell to that outside the cell, as shown in Equation 8-3.

$$E_m \propto -\frac{[K^+]_{ICF}}{[K^+]_{ECF}} \tag{8-3}$$

The symbol, $\propto$, means *proportional to*. From Equation 8-3 it may be deduced that if the ratio of K^+ concentration inside the cell to that outside the cell increases, the magnitude of E_m increases, that is, E_m becomes more negative. If the ratio decreases, then the magnitude of E_m decreases, that is, the value becomes less negative (Figure 8-4).

Both hypokalemia and hyperkalemia affect the excitability of nerve and muscle cell membranes. Excitable cells possess membranes that respond to stimuli in an all or nothing fashion, that is, they generate action potentials (Figure 8-4). To generate an action potential, the cell must first depolarize from its resting potential to a threshold potential (E_{Th}). Once the cell depolarizes to threshold, the action potential is generated automatically. Excitability is determined by the difference between resting potential and threshold as follows: the greater the magnitude of the difference, the less excitable the cell; the smaller the difference, the greater the excitability.

In hypokalemia, the K^+ concentration outside the cell (ECF) decreases and the resting potential increases from its normal value to a more negative value (Equation 8-3). Threshold potential is unaffected. If hypokalemia persists long enough for intracellular stores of K^+ to be decreased, then the ratio as shown in Equation 8-3 will be restored to normal and excitability will, therefore, also return to normal. When K^+ concentration in the ECF is less than 3 mEq/L, patients may complain of muscle weakness. The permeability of heart muscle cells to K^+ decreases with hypokalemia and causes a slowing of repolarization. This effect causes a flattening of the T wave on the electrocardiogram because repolarization is spread out over time. The slowing of repolarization increases the chances of developing arrhythmias.

Renal function is also altered with hypokalemia. There is a decreased glomerular filtration rate (GFR), reduced ability to concentrate the urine, and increased generation of bicarbonate in hypokalemia. These alterations are reversed when the hypokalemia is alleviated.

In hyperkalemia, the ratio shown in Equation 8-3 is decreased and the resulting resting membrane potential is also decreased, that is, the inside of the

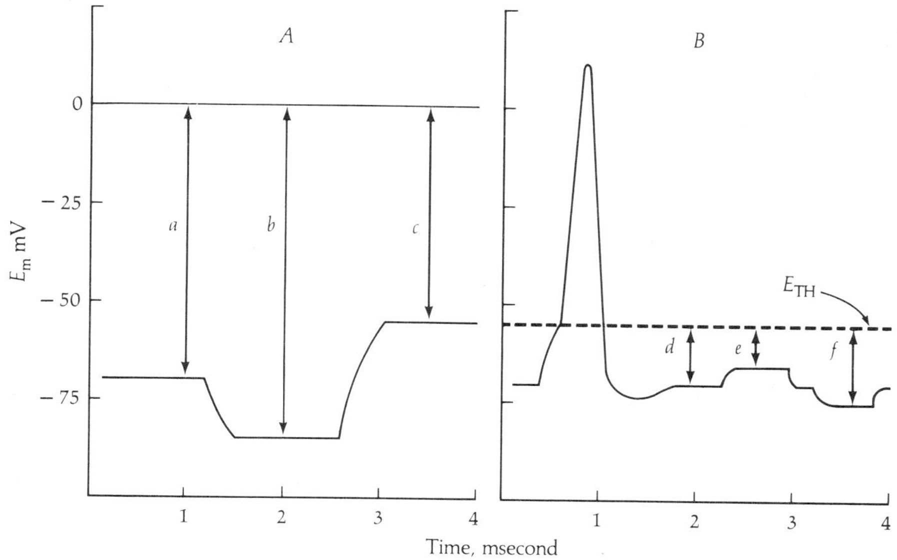

$$\text{Excitability} = \frac{1}{E_m - E_{Th}}$$

FIGURE 8-4 The effects of potassium concentration in the ECF on potential difference (E_m) across cell membranes. (*A*) Three different conditions are shown. Arrow *a* shows normal resting potential (about −70 mV). Arrow *b* shows an increased E_m and the cell is said to be hyperpolarized. Arrow *c* shows the E_m to be decreased, and the cell is said to be hypopolarized. (*B*) An excitable cell depolarizes to threshold (E_{Th} about −55 mV) and generates an action potential. After repolarization, the cell returns to resting potential and has normal excitability (arrow d), that is, the difference between E_m and E_{Th} is normal (about 15 mV). Arrow *e* shows the effects of hypopolarization due to hyperkalemia. The difference between E_m and E_{Th} is diminished and the cell is said to be more excitable. Arrow *f* shows the effects of hypokalemia, in which the cell is hyperpolarized and less excitable.

cell is less negative (Figure 8-4). This depolarization reduces the difference between the resting potential and threshold potential and makes the cell more excitable. This heightened excitability causes skeletal muscle weakness and even paralysis if the resting potential decreases below the threshold potential, because the membranes cannot repolarize. In the heart, the increased excitability can cause fatal arrhythmias.

CALCIUM

Calcium has many important functions in the body, including (1) helping to maintain structure and function of cell membranes, (2) activation of excitation–contraction coupling in muscle, (3) participation in neurotransmitter release at synapses, (4) activation of specific steps in clotting of blood, and (5) activation of complement. In addition, Ca^{++} is the principal cation involved in maintenance of bone and tooth structure and function.

Calcium Stores The body contains between 1 and 1.4 kg of Ca^{++}, most of which is located in bone. The skeleton serves as both a source and a sink for Ca^{++} in the body fluids. Less than 1% of total body Ca^{++} is found in extracellular fluid. On the average, plasma contains a total of about 500 mg. The average plasma concentration is about 10 mg/dl (or 5 mEq/L). About 40% of the Ca^{++} in plasma is bound to proteins such as albumin (36%) and globulins (about 4%). Another 13 to 14% is complexed with ions such as citrate, sulfate, and phosphate. About 47% of the Ca^{++} is freely ionized. This means that only 4 to 5 mg/dl (2–2.5 mEq/L) is free (ionized) Ca^{++} that is available for chemical and biological interaction (Figure 8-5).

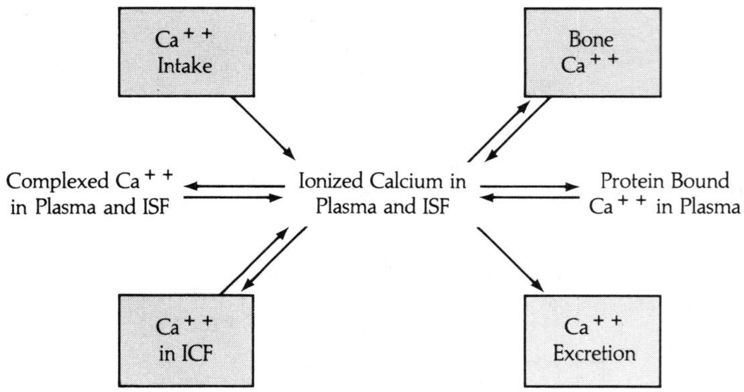

FIGURE 8-5
Calcium stores in relation to intake and excretion. Only freely ionized calcium is biologically active. The concentration of the ionized fraction depends on several complex interactions, as described in the text. Total plasma concentration is the sum of the ionized, complexed, and protein-bound calcium per liter of plasma.

Regulation of Calcium

Calcium balance is maintained by a complex set of hormonal control systems, as shown in Figure 8-6. The hormones involved include parathyroid hormone (PTH), calcitonin (CT or TCT, also called thyrocalcitonin), and vitamin D. Several organs are also involved. These include the bones, skin, liver, kidney, thyroid gland, parathyroid glands, and alimentary canal.

Intake of Calcium A person eating a balanced diet will normally ingest from 600 to 1000 mg of Ca^{++} each day. The minimum requirement is 400 to 500

FIGURE 8-6
Hormonal regulation of calcium balance. Total calcium concentration in plasma, [Ca^{++}], is closely regulated by the processes and hormones shown. CT is thyrocalcitonin (calcitonin) and PTH is parathyroid hormone. CT inhibits processes that tend to increase calcium concentration (− signs) whereas PTH increases those processes (+ signs). Vitamin D (not shown) plays a permissive role with PTH to resorb calcium from bone and increases absorption of calcium from the intestine.

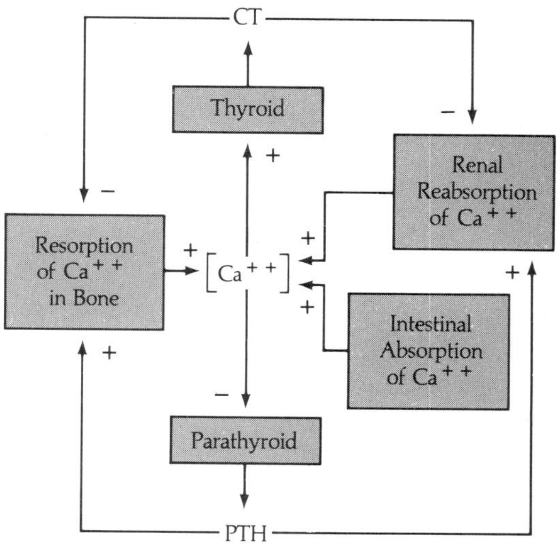

mg/day. Diets containing less than 400 mg/day will lead to negative Ca^{++} balance. Once ingested, Ca^{++} is absorbed across the intestinal epithelium into the blood. Most of the intestinal absorption of Ca^{++} occurs in the duodenum and jejunum. A smaller fraction is absorbed by the ileum. Absorption involves both diffusion and active transport, with diffusion playing a more prominent role in the ileum than in the proximal small intestine. Active transport involves a carrier protein that binds Ca^{++}.

Vitamin D plays an important role in absorption of Ca^{++}. There are several forms of vitamin D. Vitamin D_1 is a mixture of steroid compounds, all of which possess antirachitic activity. Vitamin D_2 is a plant steroid called calciferol and is a supplement added to milk. Vitamin D_3 is synthesized from a steroid (7-dehydrocholesterol) present in skin. Ultraviolet light converts 7-dehydrocholesterol to cholecalciferol (vitamin D_3). As shown in Figure 8-7, vitamin D_3 is hydroxylated in the liver to 25-OH-vitamin D_3, which also possesses antirachitic activity. In the kidney, a second hydroxyl group may be added to the number 1 position on the steroid molecule producing 1,25-$(OH)_2$-vitamin D_3, which is the most powerful of the naturally occuring forms of vitamin D (Figure 8-7). It is probable that in the intestine, vitamin D in the presence of parathyroid hormone stimulates production of the Ca^{++}-binding protein, which facilitates absorption of Ca^{++}. Vitamin D also plays a permissive role with parathyroid hormone reabsorption of Ca^{++} from bone.

Effect of PTH and CT on Calcium Balance The parathyroid and thyroid glands are both intimately involved in regulation of Ca^{++} concentration. Increasing ionized Ca^{++} concentration in plasma will stimulate release of CT from the thyroid gland and inhibit release of PTH (Figure 8-6). Calcitonin lowers Ca^{++} concentration in plasma by inhibiting resorption of Ca^{++} from

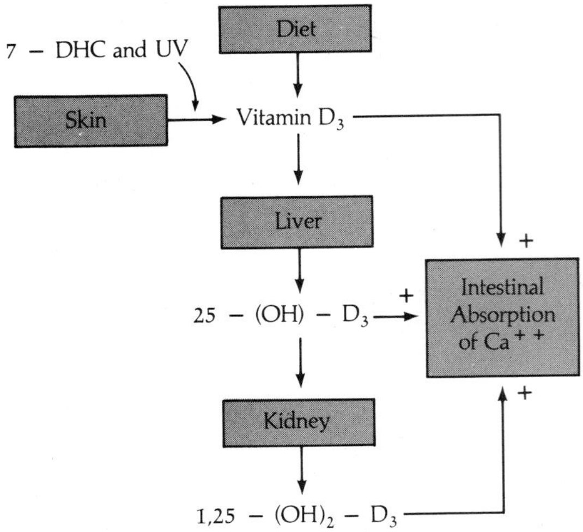

FIGURE 8-7
Vitamin D and calcium absorption. In the skin, 7-dehydrocholesterol (7-DHC) is converted in the presence of ultraviolet light to vitamin D_3. Vitamin D_3 can also be supplied in the diet. Vitamin D probably stimulates calcium absorption by increasing production of a calcium-binding protein in the intestinal mucosa.

bone (principal effect) and also increases the urinary excretion of calcium. Parathyroid hormone is secreted when ionized Ca^{++} concentration in plasma decreases. The principal effect is on bone, in which PTH stimulates osteoclasts and osteocytes to resorb bone and thereby release Ca^{++} from the matrix. In the kidney, PTH stimulates increased reabsorption of Ca^{++} in the distal tubule and enhances conversion of 25-OH-vitamin D_3 to the dihydroxy form. Parathyroid hormone does not directly affect Ca^{++} absorption by the intestine.

Excretion of Calcium Urinary excretion accounts for the bulk of Ca^{++} lost from the body each day. Ordinarily, 300 to 400 mg are lost in the urine each day. Another 150 mg can be excreted in the feces. Most of the Ca^{++} filtered in the kidney is reabsorbed. Both ionized and complexed Ca^{++} are filtered and both forms are reabsorbed. However, the complexed form is excreted preferentially to the ionized form.

Hypocalcemia

When serum Ca^{++} concentrations decrease below 8 mg/dl (4 mEq/L), hypocalcemia is present. There are four major causes of hypocalcemia in which total serum calcium concentration is decreased, as well as one category in which the concentration of the freely ionized fraction is decreased even though total serum concentration of Ca^{++} is normal (Table 8-5).

Vitamin D Deficiency As can be seen in Figure 8-7, vitamin D plays an important role in absorption of Ca^{++} from the intestine. Diets that are deficient in vitamin D can lead to hypocalcemia. Since vitamin D is a fat-soluble vitamin, diseases causing malabsorption of fats such as sprue,

Table 8-5 Causes of Hypocalcemia

I. Vitamin D deficiency	IV. Acute pancreatitis
A. Nutritional deficiency	V. Decreased ionized fraction
B. Malabsorption	A. Alkalosis
C. Disorders of vitamin D metabolism	B. Increased protein concentration
II. Thyroid and parathyroid disease	C. Chelating agents
III. Hyperphosphatemia	
A. Increased phosphate intake	
B. Renal disease	
C. Neoplasia	

chronic pancreatitis, and cirrhosis may also cause malabsorption of vitamin D. Surgical procedures that can lead to secondary malabsorption, such as partial or total gastrectomy and intestinal bypass, also cause malabsorption of vitamin D.

Specific disorders of vitamin D metabolism are also known to cause hypocalcemia. For example, vitamin D–dependent rickets is a genetic disorder in which the kidneys lack the specific enzyme to convert 25-OH-cholecalciferol to 1,25-$(OH)_2$-cholecalciferol (Figure 8-7). In chronic renal disease, loss of functional renal tissue may cause a deficiency of this enzyme. The liver may be unable to sythesize 25-OH-cholecalciferol, which also reduces the amount of vitamin D available for stimulating Ca^{++} absorption. Certain drugs metabolized by the liver such as anticonvulsants induce enzymes in liver cells, which not only metabolize the drug but also metabolize vitamin D to inactive compounds.

Disease of the Thyroid and Parathyroid Glands Inadequate secretion of parathyroid hormone diminishes resorption of Ca^{++} from bone, which is an essential part of maintaining Ca^{++} concentration in ECF. Hypoparathyroidism also reduces renal reabsorption of Ca^{++}. The net effect is hypocalcemia. Hypoparathyroidism can be caused by removal of the parathyroid glands or it may be an idiopathic disorder.

Excessive secretion of calcitonin will produce hypocalcemia because of its osteogenic effects and because it promotes reduced reabsorption of Ca^{++} in the kidney. Abnormal CT secretion occurs with medullary carcinoma of the thyroid gland and occasionally in carcinoma of the lung. It should be noted that many patients with excess CT secretion have normal serum Ca^{++} concentrations because of a secondary increase in PTH secretion.

Hyperphosphatemia In many texts, regulation of Ca^{++} and phosphate are discussed together. For pedagogical reasons phosphate is discussed separately in this text. The mechanisms of phosphate-induced hypocalcemia are presented in the section titled *Phosphate*.

Acute Pancreatitis Hypocalcemia is frequently observed in patients with acute pancreatitis. The inflammation of the pancreas causes release of proteolytic and lipolytic enzymes. It is thought that Ca^{++} combines with the fatty acids released by lipolysis to form soaps. This chemical interaction may produce the hypocalcemia associated with acute pancreatitis.

Decreased Ionized Fraction Patients with alkalosis, especially if arterial pH is greater than 7.5, may develop signs and symptoms of hypocalcemia. The total Ca^{++} concentration, however, may be normal. Calcium is bound to protein in plasma. When pH increases, more hydrogen ions dissociate from the protein, creating more negative charges on the conjugate protein base. The more negative charges available, the more Ca^{++} will combine with the protein (Equation 8-4).

$$\begin{array}{ccc} & & Ca^{++} \\ & & + \\ HPr \rightleftharpoons H^{+} + & & Pr^{-} \\ & & \updownarrow \\ & & Ca \cdot Pr \end{array} \tag{8-4}$$

Therefore, total Ca^{++} concentration can be normal, but the concentration of the ionized fraction, which is the biologically active form of Ca^{++} (Figure 8-5), will be less than normal. Signs of hypocalcemia, such as carpopedal spasm, hyperactive tendon reflexes, and symptoms such as numbness and tingling of the fingers, toes, and circumoral region, may be present. Increased protein concentration in plasma can also increase the binding of Ca^{++} and produce the same signs and symptoms as seen with decreased ionized Ca^{++} in alkalosis.

Citrate is a chelating agent added to whole blood to prevent coagulation. Citrate forms a complex with Ca^{++}, which prevents Ca^{++} from activating certain clotting factors and thereby prevents stored blood from clotting. When citrated whole blood is infused in large quantities or very rapidly, the citrate may markedly decrease the ionized Ca^{++} fraction in the blood. Calcium may have to be administered in these cases. It should be noted that in the usual situation with blood infusions the mobilization of Ca^{++} from body stores is sufficient to replace Ca^{++} chelated with citrate. It is only with large replacement volumes or very rapid infusion that Ca^{++} may need to be administered (in a separate infusion catheter) while the blood is being given.

Hypercalcemia

A wide variety of disorders are associated with or cause hypercalcemia. It is beyond the scope of this text to describe all the causes of hypercalcemia. However, those more frequently encountered in the clinical setting, as listed in Table 8-6, are discussed briefly in the following sections.

Table 8-6 Common Causes of Hypercalcemia

I. Hyperparathyroidism
II. Other endocrine disorders
  A. Hyperthyroidism
  B. Adrenal insufficiency
III. Neoplasia
IV. Excess vitamin D
V. Milk alkali syndrome

Hyperparathyroidism Hyperparathyroidism is a fairly common cause of hypercalcemia and *primary* hyperparathyroidism accounts for 15 to 20% of all incidents of hypercalcemia. The mechanism is that increased secretion of PTH mobilizes Ca^{-+} from bone and stimulates reabsorption by the distal renal tubules. The net effect is to increase the concentration of Ca^{++} in the ECF. Renal calculi are a common complication of this disorder. Surgical removal of the hypersecretory tissue is usually curative.

Other Endocrine Disorders Hyperthyroidism has been shown to be associated with hypercalcemia in about 20% of the patients with this disease. The excess thyroid hormone causes increased resorption of Ca^{++} from bone. Thyroid hormone in these cases also increases urinary excretion of Ca^{++}, which may potentiate formation of urinary calculi. Absorption of Ca^{++} from the alimentary canal also decreases; therefore, the overall effect of excess thyroid hormone on Ca^{++} balance is to mobilize Ca^{++} from bone and promote Ca^{++} loss, that is, negative Ca^{++} balance.

Patients with adrenal insufficiency frequently have hypercalcemia; however, the mechanism by which it is produced in this disease is not known.

Neoplasia The most common cause of hypercalcemia is a malignant neoplasm. First, metastases to bone frequently cause severe hypercalcemia as the bone undergoes osteolysis. Nonbone tumors may secrete substances (not PTH or vitamin D_3) that cause bone resorption. One of these substances is probably a prostaglandin. There is evidence that another yet to be identified factor from tumors stimulates osteoclast activity and increases bone resorption. In any event, resorption of Ca^{++} from bone in all of these cases can cause hypercalcemia.

Excess Vitamin D Vitamin D given in pharmacologic doses can cause hypercalcemia. In these doses vitamin D can accumulate in tissues; once hypercalcemia appears, from one to six weeks may be required before normocalcemia returns. These patients may remain normocalcemic without vitamin D therapy for several weeks after the hypercalcemia episode. It is very important to frequently measure serum Ca^{++} concentrations in patients receiving pharmacologic doses of vitamin D to ensure that hypercalcemia is not developing.

Milk Alkali Syndrome Patients with the milk alkali syndrome have hypercalcemia, hyperphosphatemia, alkalosis, metastatic calcifications, and progressive renal failure. The syndrome is caused by ingestion of large amounts of milk and calcium carbonate antacids. These patients develop hyperphosphatemia because the high serum Ca^{++} concentrations suppress PTH secretion from the parathyroid glands. (Phosphate is discussed in the next section.) If antacids other than calcium carbonate are used, such as magnesium and aluminum hydroxide mixtures, the syndrome does not develop. If the Ca^{++} load is markedly decreased, especially by cessation of calcium carbonate ingestion, the characteristics associated with the syndrome will usually be reversed. It is of interest that Ca^{++} salts cause a rebound secretion of acid by parietal cells in the stomach. It is, therefore, questionable whether Ca^{++} containing antacids should be used by patients with peptic ulcer disease.

PHOSPHATE

Stores, Intake, and Output

Phosphate is found in the extracellular fluid in both organic and inorganic compounds. Organic phosphate is present in plasma as phospholipid and is protein bound. Inorganic phosphate is found as $HPO_4^=$ and $H_2PO_4^-$ in a ratio of 4:1 at normal plasma pH. About 80% of the inorganic phosphate is filtered by the kidney and about 20% is unfilterable because it is protein bound. Most of the phosphate in the body is found in the ICF as organic and inorganic phosphate.

In the past it was thought that Ca^{++} and phosphate concentrations in ECF were reciprocally related, that is, if the concentration of one ion increased, the other decreased such that the product of the concentrations of the two was constant. It is now known that this assumption is not true and that the relationship between Ca^{++} and phosphate is more complicated than a simple reciprocal relationship. It is true that a sudden marked increase in serum concentration of phosphate will lead to a reciprocal fall in Ca^{++} concentration. However, if Ca^{++} concentration rises acutely, the phosphate concentration will not always decrease and, in fact, in some instances it too increases.

An average person ingests from 800 to 1500 mg of phosphate every day in a balanced diet. Anywhere from 50 to 65% of the ingested phosphate is absorbed from the alimentary canal. The mechanisms of regulation of phosphate absorption have not been described, but part of the absorption involves active transport. High phosphate concentration in the diet does not decrease Ca^{++}

absorption and, in fact, phosphate is required for intestinal absorption of Ca^{++}. However, phosphate absorption may decrease if the diet has high concentrations of Ca^{++}.

Phosphate is excreted in the urine and is the major urine buffer. The ratio of the monohydrogen form to the dihydrogen form depends on urine pH. The rate of excretion of phosphate depends on both oral intake and parathyroid hormone. If the dietary intake increases, urine excretion rate increases as well. Parathyroid hormone also increases the rate of renal excretion of phosphate. If oral phosphate intake decreases, urinary excretion of phosphate also decreases. High oral intake of Ca^{++} will decrease phosphate excretion either by reducing alimentary absorption or suppressing PTH secretion.

Hormonal Regulation of Phosphate

Vitamin D_3 promotes uptake of phosphate by the intestine. It also augments reabsorption of phosphate by the renal tubule and promotes resorption of phosphate from bone. In other words vitamin D_3 increases serum phosphate concentration (Figure 8-8).

Parathyroid hormone increases resorption of phosphate from bone and enhances phosphate absorption from the intestine. However, the major effect of PTH is to produce hypophosphatemia because PTH decreases phosphate reabsorption in the nephron and promotes phosphaturia (Figure 8-8).

Calcitonin inhibits resorption of phosphate from bone. It also decreases absorption of phosphate from the intestine. Finally, CT reduces reabsorption of phosphate by the kidney. The overall action of CT on phosphate is, therefore, to decrease phosphate concentration in the ECF (Figure 8-8).

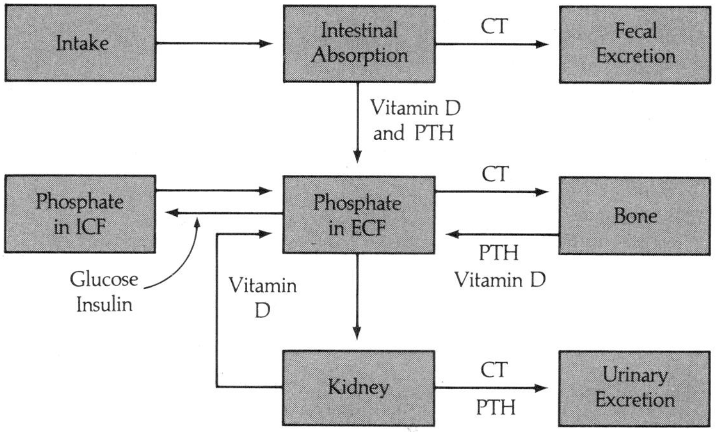

FIGURE 8-8 Processes and hormones regulating phosphate balance.

Hypophosphatemia

Phosphate concentration is usually reported in terms of the concentration of phosphorus. Normally, concentration of phosphorus in serum ranges from 2.5 to 4.0 mg/dl in adults and 4 to 7 mg/dl in children. Concentrations less than 2.5 mg/dl constitute hypophosphatemia. Total body stores of phosphate may be normal, increased, or decreased with hypophosphatemia.

Total body stores of phosphate may decrease whenever there is inadequate intake of phosphate in the diet. This is especially important for patients who have been malnourished or starved for a prolonged period of time. During starvation serum phosphorus concentrations will be normal or only mildly decreased. However, when caloric intake is again increased (either orally or with hyperalimentation), the resulting tissue growth and the concomitant cellular uptake of phosphate required for anabolism can produce severe hypophosphatemia. Aluminum and magnesium hydroxides, which are used as a part of antacid therapy in peptic ulcer disease, bind phosphate in the alimentary canal and prevent absorption of phosphate. Frequently, these patients are also hypophosphatemic from malnutrition.

Respiratory alkalosis promotes glycolysis, which causes increased uptake of phosphate by cells for phosphorylation of glucose. Therefore, hypophosphatemia may be present in patients who hyperventilate for prolonged periods (Chapter 6).

Urinary losses of phosphate increase with ketoacidosis because of the increase in titratable acidity (Chapter 4). Extra phosphate comes from the catabolism associated with diabetic ketoacidosis, which causes breakdown of organic phosphate compounds in the ICF. This phosphate is transported to the ECF, where the extracellular phosphate is lost in the urine in the formation of titratable acidity. Treatment and correction of the acidosis will stimulate tissue repair of the catabolic losses, and phosphate is then transported into the ICF from the ECF. Unless there is adequate intake of phosphate following ketoacidosis, hypophosphatemia may develop.

Chronic alcoholics may become hypophosphatemic for a variety of reasons, such as lack of adequate phosphate intake, use of antacids, and the metabolic acidosis that is associated with alcoholism.

There are several physiologic effects of hypophosphatemia. These include blood cell dysfunctions, depression of the central nervous system, resorption of bone minerals, metabolic acidosis, and cardiac myopathies. Blood cell dysfunction affects red cells, white cells, and platelets. Phosphate depletion reduces 2,3-diphosphoglycerate (2,3-DPG) in red cells, which in turn causes a decrease in the P_{50} (Chapter 3). In white cells hypophosphatemia decreases the amount of adenosine triphosphate (ATP) needed for defense against bacterial invasion. Thrombocyte count decreases in hypophosphatemia, which can lead to internal bleeding because thrombocytopenia reduces the ability of the blood to clot.

The central nervous system may exhibit dysfunction for two reasons. First, hypophosphatemia may directly alter nerve cell function, and second, the brain may become hypoxic because of the decreased P_{50} in blood caused by reduced 2,3-DPG in red blood cells.

In hypophosphatemia, bone resorption of phosphate and Ca^{++} is common. The mechanism is not fully known. Parathyroid hormone concentration probably decreases because of the hypercalcemia resulting from resorption. Synthesis of 1,25-$(OH)_2$-vitamin D_3 increases, which promotes intestinal absorption of phosphate, decreases renal excretion, and promotes resorption of phosphate from bone (Figure 8-8).

Metabolic acidosis is complicated by phosphate deficiency. With hypophosphatemia there will be a reduced titratable acidity. In addition, ammonium ion formation is also reduced in phosphate deficiency. Thus, the kidneys have a reduced ability to compensate for metabolic acidosis.

Hypophosphatemia may cause and complicate congestive heart failure. In animal studies, phosphate depletion has been shown to cause congestive heart failure. If a patient has heart failure, phosphate deficiency could decrease the amount of ATP available for contraction of the myocardium.

Hyperphosphatemia

There are three major mechanisms by which phosphate stores and concentration in the ECF are increased: increased intake, renal disease, and treatment of neoplasia. Phosphate intake may be increased orally by ingestion of potassium phosphate. Phosphate is absorbed from the intestine when phosphate-containing solutions are used in large quantities for enemas or as orally administered laxatives. Infants given milk from cows can develop hyperphosphatemia and hypocalcemia because cows' milk contains high concentrations of phosphate.

In both acute renal failure and advanced chronic renal disease, phosphate retention and mobilization of phosphate from bone can cause hyperphosphatemia. Patients who are being treated with antineoplastic agents can develop hyperphosphatemia because these agents cause lysis of the tumor cells, which in turn releases phosphate into the ECF.

Hypocalcemia is a concomitant of hyperphosphatemia. The mechanism by which Ca^{++} concentration decreases is not completely known. It has been hypothesized that the increased phosphate concentration causes increased bone formation, which would require Ca^{++}. Whatever the mechanism, the process is effective enough to use infusion of phosphate solutions to decrease Ca^{++} concentration in patients with hypercalcemia.

CHLORIDE

On the average, there are about 35 mEq of chloride per kilogram of body weight in an adult. Therefore, an 80-kg male would contain approximately

2800 mEq of chloride. Most of the chloride (about 70%) is found in the ECF, whereas the remaining 30% is distributed in the ICF and is also bound to collagen fibers in connective tissues.

Sodium and chloride are handled in a parallel fashion in the body. Thus, an individual that is hyponatremic will generally also be hypochloremic. An important example of chloride imbalance occurs in metabolic alkalosis, in which bicarbonate concentration increases at the expense of chloride concentration. Thus, hypochloremia is a common component of metabolic alkalosis. Correction of metabolic alkalosis will require replacement of chloride; otherwise the kidney will continue to generate bicarbonate to accompany reabsorption of Na^+ ions. With replacement of chloride deficits, chloride can accompany the Na^+ being reabsorbed, so that bicarbonate generation and reabsorption can be decreased to normal.

BICARBONATE

Regulation of bicarbonate concentration was discussed in Chapters 4, 5, and 6. Stores of bicarbonate ions average about 12 mEq/kg body weight or about 960 mEq in an 80-kg male. Because bicarbonate can be synthesized by the kidneys, red cells, and parietal cells in the stomach, the body is not dependent on dietary intake as a source for this anion.

MAGNESIUM

In recent years, the importance of magnesium in chemical reactions in cells and balance of this divalent cation have been recognized. The body contains about 25 g of magnesium, most of which (60%) is in bone. Approximately 20% of the total magnesium in the body is found in muscle, and of this only about 25% is exchangeable. The rest of the muscle magnesium is bound to proteins, nucleic acids, and ATP. The plasma concentration of magnesium is about 1.6 mEq/L, and 20% of that is protein bound. As was the case with calcium, only the ionized fraction is biologically active.

The regulation of magnesium balance is still not thoroughly known. Many disease processes can lead to reduction in magnesium stores, including malnutrition, malabsorption syndromes, hyperaldosteronism, diabetic ketoacidosis, primary hyperparathyroidism, and also diuretic therapy. Magnesium excess occurs in patients with adrenal insufficiency and with renal disease when patients ingest large quantities of magnesium antacids.

Hypoxia and Venous Blood-Gas Composition

CHAPTER 9

INTRODUCTION

Hypoxia is a broad general term that means that there is a decreased supply of oxygen to tissues. The implication is that the oxygen supply to systemic tissues is *less than normal*. To understand the significance of hypoxia, it is necessary to have an understanding of both how cells obtain their oxygen requirements and the reasons why cells need oxygen.

The pathway by which cells obtain oxygen is shown in Figure 9-1. Oxygen enters the lungs from the atmosphere by the process of ventilation (Chapter 4). The oxygen then diffuses across the alveolar membrane into the blood. All of the chemical interactions from addition of oxygen to blood described in Chapter 3 take place in the blood in pulmonary capillaries (see Figure 3-7). The oxygen is then circulated with the blood through the arterial tree into the systemic capillaries. Next, oxygen diffuses from capillary blood to the ISF and from the ISF into cells.

Cells need oxygen to generate energy for cell functions. Energy is required to maintain intracellular electrolyte composition and transport solutes into as well as out of ICF across the cell membrane. Energy is also used for mechanical work such as muscle contraction, metabolic conversions such as synthesis of new compounds, and conversion of lipid waste products into water-soluble compounds. Energy contained in the covalent bonds of foodstuffs is converted by the cell to high-energy phosphates in the form of ATP. It is possible to obtain some ATP from anaerobic metabolism in cytoplasm of cells, but most comes from aerobic metabolism in mitochondria (Figure 9-2). Without oxygen

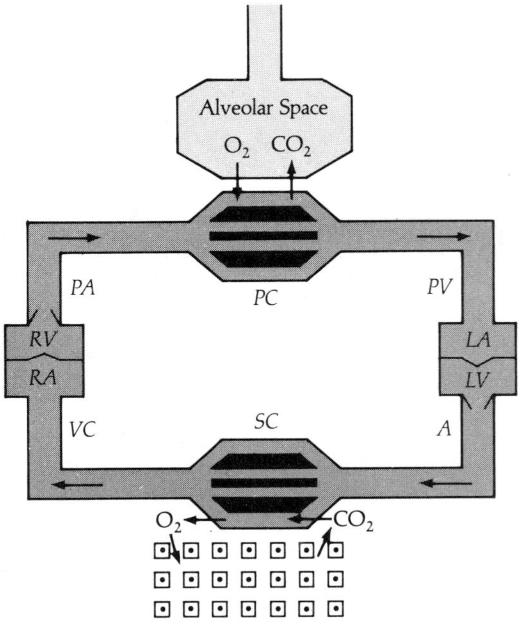

FIGURE 9-1
A model of the cardiopulmonary system illustrating transport of gases. Blood is oxygenated in pulmonary capillaries (*PC*) and flows through pulmonary veins (*PV*) into the left atrium (*LA*) and ventricle (*LV*). Blood flowing from different regions of the lungs mixes in the left ventricle to form mixed arterial blood. The blood is then ejected into the systemic arterial tree (*A*) and to systemic capillaries (*SC*). In this capillary bed O_2 diffuses out to tissue cells and CO_2 diffuses from the tissues into the blood. Blood flows from capillaries to systemic veins into the venae cavae (*VC*) to the right atrium (*RA*) and right ventricle (*RV*). Blood from the various systemic veins mixes in the right ventricle to form *mixed venous blood,* which is ejected into the pulmonary artery (*PA*).

there is insufficient energy for cells to carry out their normal functions and insufficient energy to maintain intracellular electrolyte composition. Consequently, without oxygen, cells die.

Tissues differ in their sensitivity to hypoxia. Neural tissue is the most sensitive. When aerobic metabolism is diminished in brain tissue because of hypoxia, lactic acid is produced. Because hydrogen ions and lactate cannot easily cross the blood–brain barrier, pH decreases rapidly and brain cells can be damaged. In other tissues, such as muscle, the lactic acid is more easily removed from the ISF of the tissue. Cardiac muscle is also quite sensitive to hypoxia when it is actually contracting and performing mechanical work. However, when the heart is not contracting its oxygen requirement is only about 12% of its requirement when contracting. Both hypothermia and

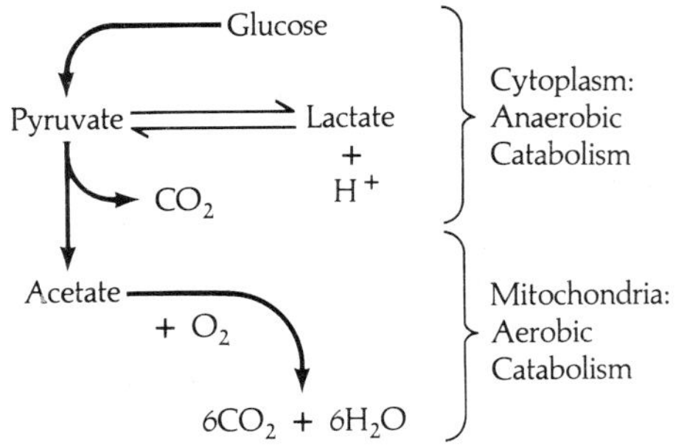

FIGURE 9-2
Pathways for metabolism of glucose used by cells to obtain energy. Oxygen is used in the metabolism in the aerobic phase, which occurs only within the mitochondria. Anaerobic metabolism takes place in the cytoplasm of cells. Two molecules of ATP are formed from each molecule of glucose during anaerobic metabolism. However, a total of 38 molecules of ATP are formed if the glucose is metabolized completely to CO_2 and H_2O. Thus, aerobic metabolism produces 19 times more ATP than is obtained from anaerobic metabolism alone. Cells that do not have mitochondria (such as red cells) cannot carry out aerobic metabolism.

anesthesia reduce oxygen requirements by decreasing metabolic rate; hence, hypothermic tissues are less sensitive tc hypoxia than normothermic tissues.

The amount of oxygen stored in the body (oxygen stores) is quite limited. When ventilation ceases, these stores are rapidly depleted and, thus, can serve to supply oxygen needs for only a brief period of time. Oxygen is stored in four places in the body: body water, myoglobin, hemoglobin, and the lungs. The body water has very little stored oxygen because this gas is very insoluble in water (0.003 ml/dl · torr at 37°C). Myoglobin in muscle is not able to supply oxygen until severe hypoxia is present. Only the lungs and hemoglobin have really useful stores of oxygen. Normally, the lungs, when at the resting midposition, have about 400 ml oxygen stored in the functional residual capacity. Another 800 to 900 ml will be stored bound to hemoglobin. If oxygen consumption averages 250 ml/minute, then only a 4 to 5-minute supply of oxygen is stored in the body. If pulmonary disease is present and the P_{O_2} of alveolar tissue is less than normal, the oxygen stores will be less than the normal supply.

When a patient breathes 100% oxygen, the O_2 stores increase to 2 to 3 L. Even under these conditions all of the oxygen stores would be used up in about 12 minutes with apnea. Therefore, since oxygen stores are quite small, oxygen used up by metabolism must be continually replenished from the atmosphere to maintain life.

CAUSES OF HYPOXIA

Hypoxia can be produced by a variety of disease processes. Generally, the causes may be classified under three major headings: (1) decreased oxygen supply in arterial blood, (2) decreased cardiac output, and (3) decreased oxygen utilization. Each of these is discussed below.

Decreased Oxygen Supply in Arterial Blood: Arterial Hypoxemia

A decreased supply of oxygen in blood is called *hypoxemia*. Causes of arterial hypoxemia are shown in Table 9-1. The first five causes listed produce both a reduced arterial P_{O_2} and a reduced oxygen concentration in arterial blood (decreased oxygen content). With anemia, the arterial P_{O_2} may be normal, but oxygen content will be reduced.

Decreased $P_{I_{O_2}}$ When the inspired P_{O_2} ($P_{I_{O_2}}$) is reduced, both alveolar and arterial P_{O_2} will decrease. A reduction in the inspired P_{O_2} can occur (1) when oxygen is removed experimentally from the inspired gas, (2) when oxygen is consumed by oxidation, as occurs in fires, (3) when people are entrapped in closed spaces, and (4) when an individual breathes air at high altitudes. The net effect is identical in all four situations—inspired P_{O_2} decreases. At high

Table 9-1 Causes of Hypoxia

I. Arterial hypoxemia	II. Decreased blood flow (ischemic hypoxia)
A. Decreased inspired P_{O_2}	A. Local / regional
B. Impaired diffusion across the alveolar membrane	B. Generalized systemic
C. Hypoventilation	1. Heart failure
D. Right to left shunts	2. Hypovolemia
E. Mismatching ventilation and perfusion	III. Decreased oxygen utilization
F. Anemia (anemic hypoxia)	A. Low P_{50}
	B. Enzyme poisons (histotoxic hypoxia)

altitude, the inspired P_{O_2} decreases because barometric pressure decreases (Figure 9-3 and Equation 9-1).

$$P_{I_{O_2}} = 0.2095\,(P_B - 47) \tag{9-1}$$

In Equation 9-1, P_B is barometric pressure, the number 47 is the vapor pressure of water, 0.2095 is the fractional concentration of oxygen in air, and $P_{I_{O_2}}$ is the inspired P_{O_2} (see Chapter 3).

As described in Chapter 3, the percentage of hemoglobin saturated with oxygen depends on the P_{O_2} of the blood. In pulmonary capillaries, blood ordinarily comes into equilibrium with gas in the alveolar space such that the P_{O_2} of blood leaving the pulmonary capillaries is equal to the P_{O_2} of alveolar gas. In this sense, then, the percentage of hemoglobin saturated with oxygen is determined by alveolar P_{O_2}. However, the rate of diffusion of oxygen into blood from alveoli depends on the magnitude of the difference in the P_{O_2} between alveolar gas and blood in pulmonary capillaries. When alveolar P_{O_2} decreases, this difference decreases. For example, in healthy people at sea level, the alveolar P_{O_2} is approximately 100 torr whereas the P_{O_2} of blood entering pulmonary capillaries (mixed venous blood) is about 40 torr. The difference in

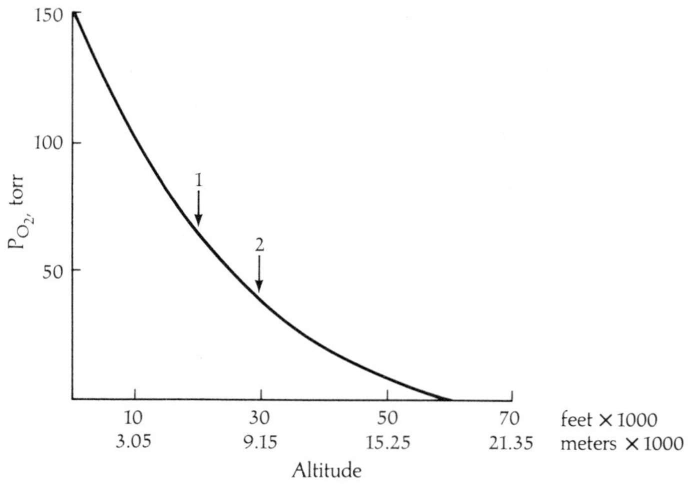

FIGURE 9-3
The relationship between inspired P_{O_2} and altitude. The values shown on the abscissa are multiplied by 1000. Arrow 1 shows the inspired P_{O_2} at the top of Mt. McKinley (Alaska) and arrow 2 the inspired P_{O_2} at the top of Mt. Everest (Tibet). The alveolar P_{O_2} may be calculated from the inspired P_{O_2} by subtracting the value of the alveolar P_{CO_2}/R from the $P_{I_{O_2}}$ on the graph.

FIGURE 9-4
Time course for oxygenation of hemoglobin: *pa* = time in pulmonary artery; *pv* = time in pulmonary veins; curve *a* shows the change in P_{O_2} in a red cell for people with normal alveolar membranes; curve *b* shows the change in P_{O_2} in a red cell for people with a thickened alveolar membrane or reduced surface area in the lung; *c* is the duration of time that a red cell is in the pulmonary capillaries at normal resting cardiac output; *d* is the duration of time a red cell is in the pulmonary capillaries with maximum cardiac output (during exercise); *e* is the P_{O_2} difference between alveolar gas and blood entering pulmonary capillaries and represents the driving pressure to oxygenate the blood. (*A*) Relationships for people breathing air at sea level; (*B*) relationships for people breathing air at high altitude. The time course for saturation is longer in *B* even for normal alveolar membranes because the P_{O_2} gradient is decreased from that at sea level. The dashed horizontal lines represent alveolar P_{O_2}. Note also that at higher altitudes, even people with normal alveolar membranes may not achieve equilibrium with alveolar P_{O_2} during exercise. Hence, at higher altitudes a normal person may have a significant alveolar – arterial P_{O_2} difference. Modified from *Respiratory Pathophysiology*, by M. Kryger. Copyright © 1981 by John Wiley & Sons, Inc. Reprinted by permission.

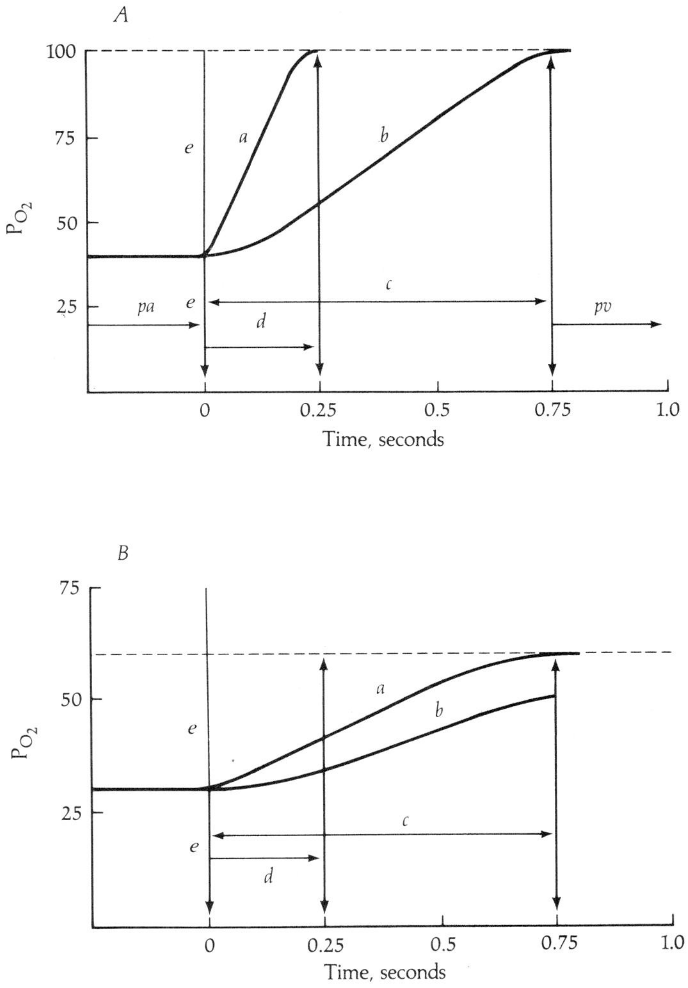

P_{O_2} between alveolar gas and pulmonary capillary blood under these conditions is about 60 torr. If these healthy people ascend to an altitude of 13,000 feet (3.96 km), the alveolar P_{O_2} would be reduced to about 60 torr whereas the P_{O_2} of mixed venous blood would be decreased to 30 torr. The P_{O_2} difference between alveolar gas and blood entering the pulmonary capillaries at this altitude would be only 30 torr. Thus, the partial pressure difference at 13,000 feet is only one half of its value at sea level. Not only will the P_{O_2} of arterial blood be decreased because of the decrement in the inspired P_{O_2}, but also the rate of diffusion of oxygen into blood will be reduced because of the decreased partial pressure difference. As a result, the time required to equilibrate the P_{O_2} of blood with that of alveolar gas will be increased (Figure 9-4). The net effect of decreased inspired P_{O_2} on arterial blood is decreased arterial P_{O_2} and decreased oxygen content—in other words, arterial hypoxemia.

Impaired Diffusion Another cause of hypoxemia is impaired diffusion of oxygen across the alveolar membrane. Impaired diffusion occurs when the total surface area of the membrane decreases or when the thickness of the membrane increases (Chapter 1). Total surface area of the alveolar membrane is decreased with emphysema or partial pneumonectomy. On the other hand, the membrane is thickened in the presence of pulmonary or alveolar edema. It should be noted that, at rest, impaired diffusion is seldom a cause of hypoxemia. However, with exercise or when alveolar P_{O_2} is decreased, impaired diffusion may make a significant contribution to hypoxemia (Figure 9-4).

Hypoventilation Hypoventilation is a third cause of hypoxemia. By definition, hypoventilation is said to be present when alveolar P_{CO_2} is increased above normal. Hypoventilation implies that there is an inadequate volume of fresh air entering the alveolar space. The net effect of hypoventilation is increased P_{CO_2} and reduced P_{O_2} of arterial blood. The hypoxemia can be corrected by enriching the inspired air with additional oxygen or by improving ventilation rate. Hypoventilation occurs with emphysema, severe pneumonia, depression of the respiratory control centers, and weakened respiratory muscles (Table 6-2). It should also be remembered that the increased P_{CO_2} accompanying hypoventilation leads to respiratory acidosis. The pathophysiology of respiratory acidosis is discussed in Chapter 6.

Shunts Shunts can cause hypoxemia. A shunt is a communication between two parts of the vascular system that permits blood to bypass a site of capillary exchange. Three categories of shunts occur in the body.

The first type is anatomically normal shunts within the systemic vascular system. For example, blood may be shunted directly from an arteriole to a venule in a microcirculatory bed through an arteriovenous anastomosis or in skin capillary beds through preferential channels and metarterioles (see Figure 2-2).

A second type, the left to right shunt, is abnormal and is found in the heart and great vessels. In these shunts, blood flows from the higher pressure found in the left side of the heart and aorta back to the right side of the heart and pulmonary artery. Examples include atrial septal defects, ventricular septal defects, and patent ductus arteriosus (Figure 9-5). Left to right shunts do not cause hypoxemia because the shunted blood is already oxygenated and is flowing back to the right side of the heart to be recirculated through pulmonary capillaries. The high pressure does enlarge the right side of the heart, causing right-sided hypertrophy and also pulmonary hypertension.

A third type of shunt, the right to left shunt, causes mixed venous blood to enter the systemic arterial system without being oxygenated. These shunts can occur with atrial and ventricular septal defects when pressure on the right side of the heart exceeds that on the left (Figure 9-5). These shunts also are produced when blood flows to alveoli that are not being ventilated, for example, in patients with atelectasis, lobar pneumonia, and pneumothorax. In these latter examples, the blood vessels involved would normally not be

FIGURE 9-5
Models of two kinds of shunts. (*A*) A model of left to right shunts. These shunts permit oxygenated blood to flow from the postcapillary side of the pulmonary circulation to the precapillary side. Shunt 1 is typical of a patent ductus arteriosus and 2 represents shunts between the left and right sides of the heart. These types of shunts do not produce cyanosis but do increase the workload of the right heart and can lead to heart failure. *SVC* = superior vena cava; *IVC* = inferior vena cava; *PA* = pulmonary artery. (*B*) A model of a right to left shunt is shown. Blood flows from precapillary vessels through the shunt, *S*, bypassing the alveolar exchange surface. Shunted blood does not participate in gas exchange with alveolar gas. Right to left shunts can occur when alveoli are not ventilated (Figure 9-6) or if the flow reverses in the shunts shown in *A*.

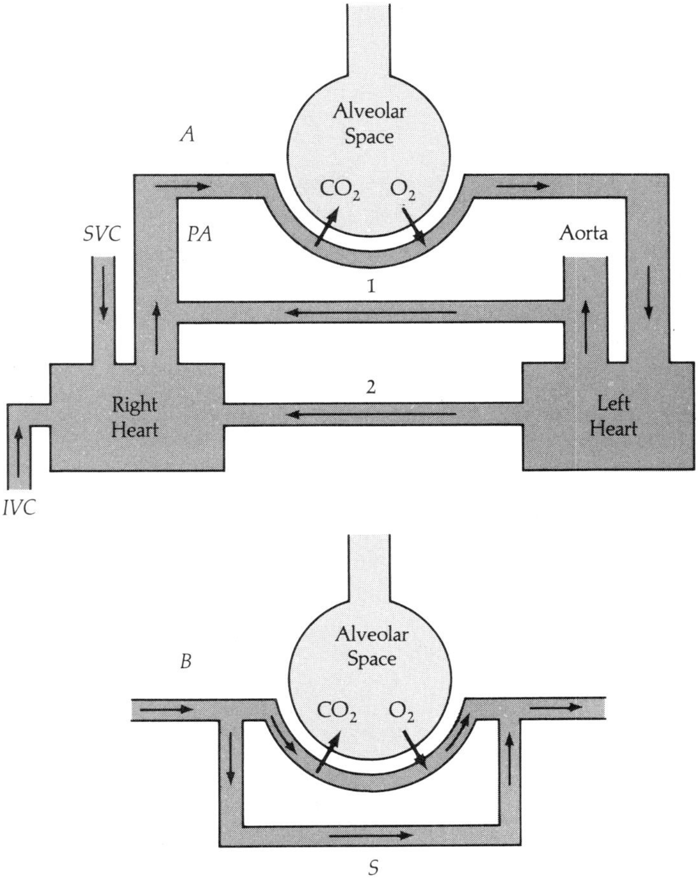

considered shunts; however, because gas exchange is prevented by lack of ventilation in these situations, mixed venous blood does not become oxygenated and is shunted to the left ventricle (Figure 9-6).

The degree of hypoxemia produced with right to left shunt depends on the fraction of blood flow bypassing exchange in pulmonary vessels (percentage of blood shunted) and on the amount of oxygen in mixed venous blood. The greater the volume of blood flowing through the shunt, the lower will be the P_{O_2} of the mixed arterial blood (Figure 9-7). Furthermore, for a given fraction or percentage of blood shunted, the less oxygen contained in mixed venous blood, the lower will be the resulting oxygen content and P_{O_2} of mixed arterial blood (Figure 9-7). Therefore, any situation that causes decreased mixed venous oxygen content, such as anemia, reduced cardiac output, or increased metabolic activity, will accentuate the arterial hypoxemia produced by a shunt.

If the shunt fraction is low, the arterial P_{CO_2} may be normal even though P_{O_2} is reduced significantly. The central chemoreceptors will cause hyperventilation of patent alveoli, thereby reducing P_{CO_2} in blood perfusing well-ventilated regions of the lung. The resulting mixture of shunted blood with high P_{CO_2} and

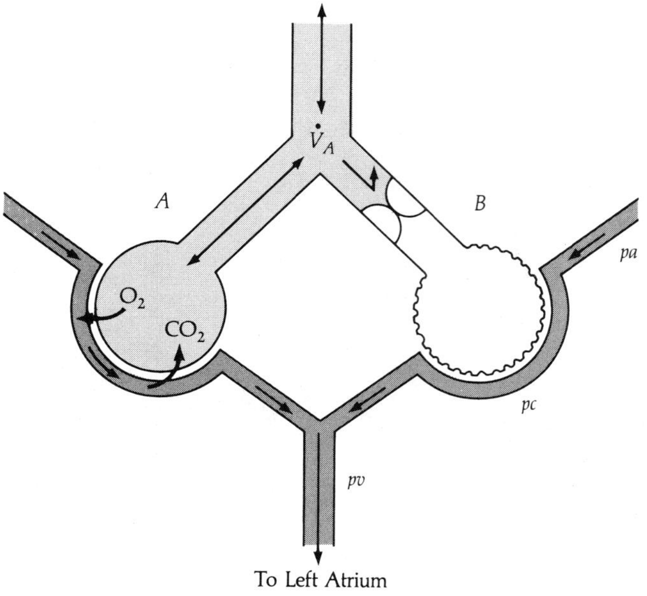

FIGURE 9-6
A model of pulmonary blood flow to unventilated alveoli. Alveoli in section *A* are well ventilated and normal gas exchange occurs across alveolar membranes in this section. The alveoli in section *B* are not ventilated, consequently the gas composition of those alveoli is essentially the same as mixed venous blood. No net gas exchange will occur in section *B*. Therefore, blood flow to alveoli in section *B* constitutes a right to left shunt. *pa* is a branch of the pulmonary artery; *pc*, the pulmonary capillaries; and *pv*, pulmonary veins.

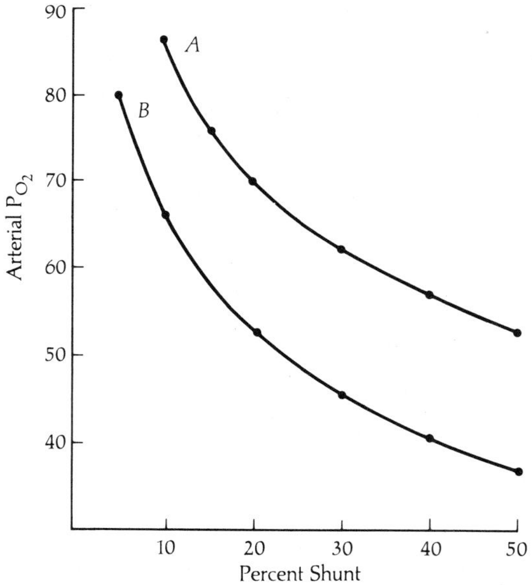

FIGURE 9-7
Change in arterial P_{O_2} as a function of the percentage of cardiac output that flows through a right to left shunt. These curves were determined from calculations in which pulmonary capillary blood was assumed to be 97.5% saturated and mixed venous blood was (*A*) 75% saturated and (*B*) 40% saturated. Hemoglobin concentration was assumed to be 15 g/dl. The decreased saturation assumed for curve *B* could be produced by anemia, decreased cardiac output, or arterial hypoxemia itself.

pulmonary capillary blood with low P_{CO_2} will have a normal P_{CO_2}. However, because the hemoglobin in blood leaving well-ventilated alveoli is nearly saturated, increasing ventilation to those alveoli (Figure 9-6*A*) will not markedly increase the oxygen content of that blood. Consequently, hyperventilation of well-ventilated alveoli will not offset the dilution of oxygen produced by the shunt.

Providing oxygen therapy for patients with shunts will increase P_{O_2} of arterial blood if the shunt is small. However, for larger shunts (greater than 25–30%) even if the patient inspires 100% oxygen, the P_{O_2} of arterial blood will be increased only marginally. The reason for such a small response to oxygen therapy in cases of larger shunts is that the added oxygen enters only that blood already saturated with oxygen, and therefore increases only the volume of physically dissolved gas. The added oxygen cannot enter the shunted blood. Once hemoglobin is saturated, breathing 100% oxygen will add only about 2 ml of O_2 per deciliter of blood. Hence, with a 50% shunt, inspiring pure oxygen will increase the P_{O_2} of arterial blood by only 5 to 10 torr.

When the shunt is caused by a decreased alveolar exchange surface due to atelectasis or pneumothorax, positive pressure ventilation will frequently help inflate collapsed alveoli and decrease the magnitude of the shunt. If the effects of the shunt are magnified by low cardiac output or anemia (curve *B*, Figure 9-7), then improving the cardiac output and correcting the anemia will help improve systemic tissue oxygenation. It should be remembered that the ultimate focus of the therapy should be to improve tissue oxygenation, not just arterial P_{O_2}. For example, positive pressure ventilation may open atelectatic regions of the lungs but may reduce cardiac output as well. If the cardiac output decreases sufficiently, tissue oxygenation may not improve even though arterial hypoxemia is diminished. This point is discussed further in subsequent sections of this chapter. A summary of the calculation of shunt fraction is presented in Appendix 2.

Mismatching Ventilation and Perfusion The most common cause of arterial hypoxemia in lung disease is mismatching of ventilation and perfusion. This topic was discussed in Chapter 4. When alveoli receive too little ventilation for the blood flow perfusing their pulmonary capillaries, the blood leaving those capillaries will be hypoxic. This situation is similar to a shunt shown in Figure 9-6 but differs in that ventilation of the affected alveoli is greater than zero. A model of mismatching ventilation and perfusion is shown in Figure 9-8. Blood leaving poorly ventilated alveoli in Figure 9-8 will have a lower P_{O_2} than blood leaving well-ventilated alveoli. If oxygen therapy is administered to patients that have hypoxemia from mismatching of ventilation and perfusion, the P_{O_2} of both well-ventilated *and poorly ventilated* alveoli will increase because oxygen can enter both kinds of alveoli. Therefore, unlike cases of shunts, oxygen therapy will increase oxygenation of arterial blood even with marked mismatching of ventilation and perfusion.

Anemia Recall from Chapter 3 that oxygen is carried in two forms in blood: (1) as physically dissolved gas and (2) as oxyhemoglobin. When the hemoglobin concentration in blood increases, then more oxygen can be carried in the blood and vice versa (see Chapter 3, Figure 3-5). In anemia the hemoglobin concentration is decreased and, therefore, the total amount of oxygen that can be carried at a given P_{O_2} is also reduced. For example, if the arterial P_{O_2} is 100

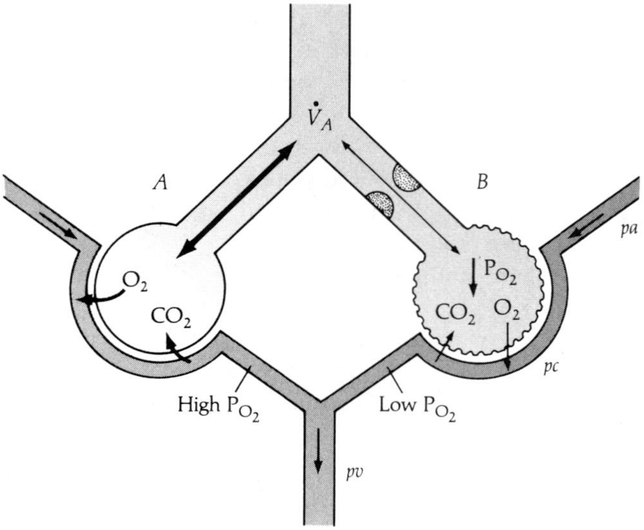

FIGURE 9-8
A model for mismatched ventilation and perfusion. Lung units labeled *B* receive too little ventilation for the amount of blood flow they receive. The blood leaving section *B* will be hypoxic compared to that of *A*. The resulting mixture of blood will also be hypoxic. This is a very similar situation to that seen with shunts but differs in that O_2 therapy will improve the P_{O_2} of alveoli in section *B* and the blood perfusing those alveoli.

torr and hemoglobin concentration is 15 g/dl, arterial blood would contain 20.6 ml oxygen/dl blood at a P_{O_2} of 100 torr (97.5% saturation) and a temperature of 37°C. If, under the same conditions, an individual were anemic with a hemoglobin concentration of only 10 g/dl of blood, then the oxygen concentration would be only 13.9 ml oxygen/dl of arterial blood (Figure 9-9). Thus, it is apparent that even if arterial P_{O_2} is normal, hypoxemia of arterial blood will occur if hemoglobin concentration is less than normal.

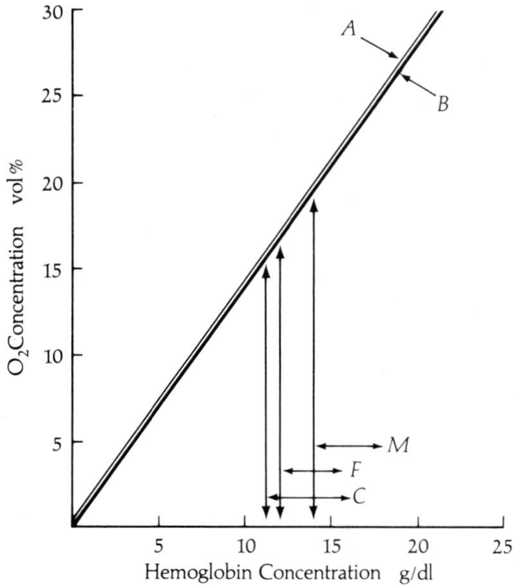

FIGURE 9-9
The concentration of oxygen in arterial blood as a function of hemoglobin concentration. The lines were determined by assuming a temperature of 37°C, pH = 7.4, a P_{O_2} of 100 torr, and a 97.5% saturation of hemoglobin. Line *A* is total oxygen concentration, that is, physically dissolved oxygen plus oxygen combined with hemoglobin. Line *B* is the oxygen combined with hemoglobin only. Ranges for hemoglobin concentration are shown by the horizontal arrows. *M* shows the range for adult males, *F*, the range for adult females, and *C*, the range for children. The vertical arrows indicate the minimum normal hemoglobin concentration for each age group and the corresponding minimum concentration of O_2 carried when conditions are as specified. Values to the left of the arrows constitute anemia for each age group.

Table 9-2 Causes of Local / Regional Ischemia

1.	Atherosclerosis
2.	Stenosis of blood vessels
3.	Vasospasm (drugs, Raynaud's disease)
4.	Embolization (thrombus, fat, or air)

Decreased Blood Flow to Systemic Tissues

Ordinarily blood undergoes gas exchange in pulmonary capillaries and oxygen is loaded into the blood at this point. Even with no arterial hypoxemia, tissues may not receive an adequate supply of oxygen. The blood that contains the oxygen must circulate through systemic tissue capillaries so that oxygen can leave the blood and diffuse into ISF and cells. Any processes that reduce blood flow to tissues (ischemia) can also cause tissue hypoxia.

There are two broad categories of ischemic hypoxia: (1) local (or regional) ischemic hypoxia and (2) generalized systemic ischemic hypoxia. The local or regional type is caused by reduced blood flow to a specific organ or region, usually because of increased resistance to blood flow. For example, atherosclerosis of the iliac arteries can increase resistance to blood flow to the lower extremities. As long as oxygen consumption by the muscles of that region remains sufficiently low, tissue hypoxia will not be an apparent problem. However, exercise (even walking) may increase oxygen consumption beyond what can be supplied by blood flow to the region. The resulting ischemic hypoxia will limit the activity of the tissues in the affected region.

It should be noted that atherosclerosis can affect any of the larger arteries of the body. In addition to the iliac arteries, the common carotid, celiac, mesenteric, renal, and coronary arteries are also vulnerable to atherosclerosis. Consequently, tissues and organs supplied by those arteries can become subject to hypoxia from ischemia produced by atherosclerosis. Other causes of regional ischemia and hypoxia are listed in Table 9-2. Any of these processes can reduce blood flow to the point that the ischemic tissues are hypoxic. Severe hypoxia can also lead to infarction of the affected tissues.

Generalized ischemic hypoxia can be caused by either heart failure or hypovolemia. When hypoxia is caused by either of these conditions, the heart pump is unable to maintain an adequate blood flow and hence is unable to maintain oxygen delivery to systemic tissues. The arterial blood in both cases may not be hypoxemic but tissues will still be hypoxic because of inadequate oxygen delivery. If hypoxemia is present with inadequate blood flow, the tissue hypoxia will be worse than with ischemia alone. Therefore, it is essential to maintain an adequate arterial P_{O_2} in the presence of heart failure or hypovolemia. Treatment of generalized ischemic hypoxia consists of improving myocardial function or increasing blood volume to maintain an adequate circulation of blood.

Decreased Oxygen Utilization

Two factors can reduce oxygen uptake by tissues once the oxygenated blood is in the systemic capillaries: (1) decreased P_{50} of arterial blood and (2) enzyme poisons. When the P_{50} of hemoglobin is decreased, the P_{O_2} of tissues must also decrease in order to unload the oxygen from hemoglobin (see Chapter 3). A decrease in P_{50} occurs with alkalemia (Chapter 6), decreased temperature of blood, decreased 2,3-DPG concentration, and carbon monoxide poisoning.

Enzyme poisons, especially cyanide, block the cytochrome enzymes in mitochondria and inhibit oxidative phosphorylation. The net effect is that mitochondria are unable to use the oxygen that is delivered, and hence cells must resort exclusively to anaerobic metabolism to obtain energy. In the latter case, the cells behave as though they were hypoxic in the presence of an adequate supply of oxygen.

RESPONSES TO HYPOXIA

Four systems are strongly affected by hypoxia: the respiratory, cardiovascular, hematopoietic, and nervous systems. The first three can generate compensatory responses to hypoxia whereas the nervous system exhibits marked dysfunction as hypoxia progresses. The compensatory responses, which are discussed below, can themselves lead to severe dysfunction. The responses may, therefore, be maladaptive.

Respiratory Responses

Hypoxemia of arterial blood stimulates the carotid and aortic body chemoreceptors. These, in turn, stimulate the respiratory neurons in the medulla to increase alveolar ventilation rate and, thereby, decrease alveolar P_{CO_2} (Chapter 4). Decreasing alveolar P_{CO_2} will cause alveolar P_{O_2} to increase, as shown in Equation 9-2.

$$P_{A_{O_2}} = P_{I_{O_2}} - P_{A_{CO_2}}/R \qquad (9\text{-}2)$$

$P_{A_{O_2}}$ is alveolar P_{O_2}, $P_{A_{CO_2}}$ is alveolar P_{CO_2}, and R is the respiratory exchange ratio. This equation is explained further in Appendix 3. When the P_{CO_2} of alveolar gas is decreased, the alveolar P_{O_2} approaches the value of the inspired P_{O_2}. Hypoxia of neural tissue will also reduce the respiratory response to hypoxemia. In instances of marked hypoxemia, the major drive to the respiratory centers in the brain may come from the carotid and aortic bodies.

Cardiovascular Response

There are two components to the cardiovascular response to hypoxia. First, cardiac output increases due to increased sympathetic stimulation. This response tends to offset the effects of arterial hypoxemia by increasing delivery of arterial blood to systemic tissues. If the hypoxemia lasts for more than a few days, this particular part of the cardiovascular response becomes blunted. Second, there is a redistribution of blood flow such that both brain and myocardium receive an increase in blood flow out of proportion to the increase in cardiac output generally. Nonetheless, the arterial hypoxemia will still cause hypoxia of neural tissue.

Hypoxia produced by decreased inspired P_{O_2} will cause constriction of pulmonary arterioles. This response is part of the normal mechanism by which pulmonary blood flow is distributed to regions that are better ventilated and the lung thereby maintains a better degree of matching ventilation and perfusion. However, when the hypoxia is generalized to all airways as occurs at high altitude, there is a vasoconstriction of the majority of pulmonary arterioles, which in turn causes pulmonary hypertension (Figure 9-10). The pulmonary hypertension increases afterload of the right heart and, if not relieved, can lead to right heart failure. Some individuals who are hypoxic because of exposure to high altitude develop a syndrome called mountain sickness. Pulmonary edema can be one of the components of this syndrome. Evidently not all of the pulmonary arterioles are uniformly constricted and there is damage from the high pressure and turbulent flow to pulmonary capillaries downstream in those capillary beds supplied by less constricted arterioles. If not severe, the damage may be reversed by supplementing the inspired air with oxygen and taking the individual to lower altitudes.

FIGURE 9-10
Mean pressure in the pulmonary artery as a function of alveolar P_{O_2}. Note that the change in pressure is more pronounced when alveolar P_{O_2} is less than 80 torr. The curve shown is a mean value and there is considerable variation about the mean.

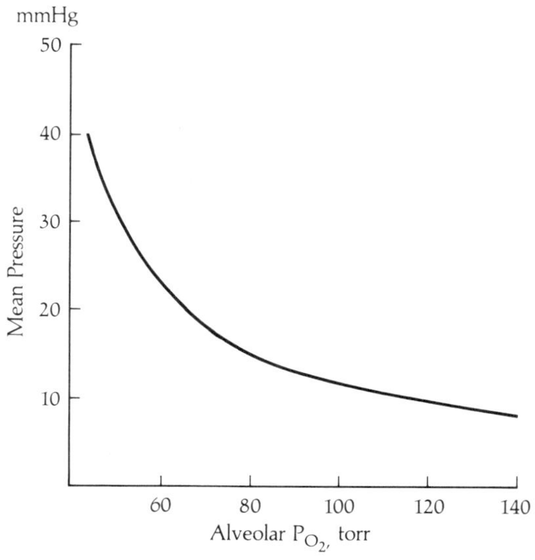

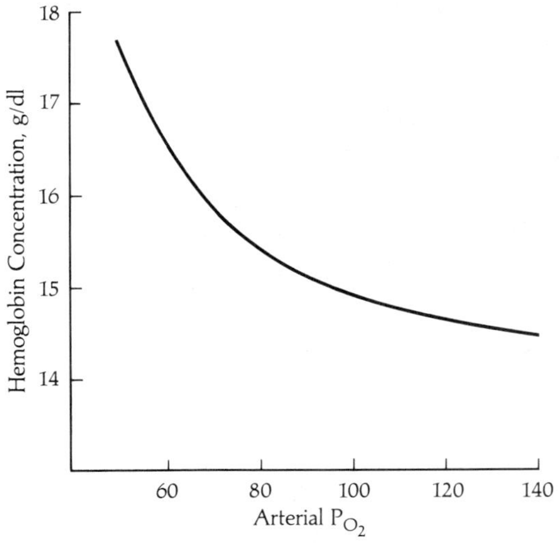

FIGURE 9-11
The effect of hypoxemia on hemoglobin concentration in blood. Hypoxia causes production of erythropoietin, which in turn stimulates red blood cell production in red bone marrow. The resulting erythrocytosis is reflected as an increase in the hemoglobin concentration. The curve shown represents the mean hemoglobin concentration. Actual values can vary considerably around this mean value.

Hematopoietic Responses

In hypoxia, a substance is released from the kidneys (90–95%) and liver (5–10%) that causes production of erythropoietin. Erythropoietin stimulates bone marrow to produce more red blood cells. The result is an increase in red cell mass (polycythemia or erythrocytosis), which provides a greater hemoglobin concentration in blood for transporting oxygen (Figure 9-11). The negative side of this compensatory response is an increased workload for the heart because polycythemia causes the viscosity of the blood to increase. Increased blood viscosity aggravates ventricular failure. Furthermore, the resistance to flow increases in direct proportion to increases in the viscosity. As a result, blood flow to the brain can decrease and reduce oxygen delivery when viscosity becomes too high.

Response of the Nervous System

Nervous tissue is very sensitive to hypoxia. With mild hypoxia, the dysfunction is reversible; however, severe hypoxia causes death of nerve cells. In general few if any effects will be noted until arterial P_{O_2} is less than 80 torr. When arterial P_{O_2} ranges between approximately 45 and 80 torr, there is a reduction in visual sensitivity. The rods in the retina are very sensitive to hypoxia and a detectable decrement in night vision occurs when arterial P_{O_2} is reduced to about 70 torr (94% saturation of hemoglobin). This P_{O_2} will occur in arterial blood of healthy people at an altitude of about 6500 feet. When arterial P_{O_2} is only 50 torr the visual fields decrease in size, the blind spot enlarges, and there

Table 9-3 A Comparison of the Normal Composition of Arterial and Mixed Venous Blood

	Arterial	Mixed venous
pH	7.40	7.38
P_{CO_2}, torr	40	46
HCO_3^-, mEq / L	24	26
P_{O_2}, torr	100	40
Percent saturation	97	75
O_2 concentration, ml / dl	20	15

is some loss of color discrimination. When arterial P_{O_2} is between 35 and 50 torr, there is impaired coordination and markedly reduced ability to perform mathematical calculations. When arterial P_{O_2} is between about 25 and 32 torr, unconsciousness will occur in only a few minutes, and if the P_{O_2} is less than 20 torr for a few minutes, irreversible damage to nerve cells occurs. In general, nerve cell function is depressed by hypoxia.

ASSESSMENT OF HYPOXIA

Arterial blood represents the output from the lungs and, after being mixed in the left ventricle, is pumped throughout the body to all systemic tissues (Figure 9-1). The blood-gas composition of arterial blood reflects the results of exchange between pulmonary capillary blood and alveolar gas, that is, arterial blood-gas composition reflects the balance between alveolar ventilation and pulmonary capillary blood flow. When the P_{O_2} of arterial blood is decreased below about 80 torr, arterial hypoxemia exists. If hemoglobin concentration is less than normal, arterial hypoxemia (in terms of total oxygen concentration) is probably present, even if the arterial P_{O_2} is normal (Figure 9-9). Normal values of arterial P_{O_2} and oxygen content are shown in Table 9-3.

When cyanosis is present, hypoxia is present. Cyanosis is a bluish discoloration of the mucous membranes, nail beds, skin, and conjunctivae of the eyes. Its presence indicates hypoxia in the affected tissues. When cyanosis is present only in the peripheral extremities, for example, the fingers or toes, but the buccal (oral) mucosa is free of cyanosis, then the regional cyanosis is most likely due to a reduced blood flow to that tissue such as occurs in Raynaud's disease, shock, or exposure to cold. However, if the oral mucosa and conjunctivae are also cyanotic (central cyanosis), the arterial hypoxemia is probably caused by one of the processes listed under Roman numeral I in Table 9-1 or by heart failure, and all systemic tissues are hypoxic to some degree.

It must be remembered that cyanosis is not an accurate indicator of either P_{O_2} or oxygen content of arterial blood. For cyanosis to be present, about 5 g of hemoglobin per deciliter of blood must be desaturated (or reduced). An individual could have erythrocytosis and have cyanosis of some tissues but an adequate oxygen supply to those tissues. Another person who is anemic may not show signs of cyanosis because he or she has less than 5 g of desaturated hemoglobin/dl of blood, but still have rather profound tissue hypoxia. Therefore, to properly evaluate the extent or degree of arterial hypoxemia, a blood-gas analysis of an arterial blood sample must be done.

Up to this point the discussion of assessment of hypoxia has been focused on arterial hypoxemia. However, tissue hypoxia can be present even if there is normal gas exchange in the lungs and arterial blood is not hypoxemic. Reduced cardiac output in heart failure or volume contraction are examples of decreased blood flow or ischemia-related causes of tissue hypoxia. It is true that if arterial blood is hypoxemic, systemic tissues are probably hypoxic as well, but lack of arterial hypoxemia does not ensure lack of tissue hypoxia. If blood flow through systemic tissues is reduced even when arterial blood is normally oxygenated, oxygen delivery to those tissues will also be reduced. It is reasonable, therefore, to conclude that tissue hypoxia cannot always be assessed from measurements of arterial P_{O_2} alone. Other assessment procedures, such as those dealing with cardiovascular status that are designed to determine the adequacy of tissue perfusion, will help in evaluation of tissue oxygenation.

VENOUS BLOOD-GAS COMPOSITION

Venous blood, especially mixed venous blood, is being used with increasing frequency in the clinical setting for assessment. The discussion of venous blood-gases is presented in this section to clarify the concepts of acid-base and oxygenation status of tissues as well as the assessment of both.

General Determinants of Venous Blood-Gas Composition

Venous blood, in general, has a blood-gas composition that reflects the composition of the ISF of the tissues from which it drains. For example, pulmonary venous blood has essentially the same composition as that found in alveoli. Each region of the lung has its own degree of matching ventilation and perfusion; hence, blood from one region of the lungs will not have precisely the same composition as blood from another region. However, blood from all regions of the lungs mixes in the left ventricle to become mixed arterial blood, which is uniform in composition. Therefore, systemic arterial blood is really mixed pulmonary venous blood that has been ejected into the arterial vessels by the left ventricle. The blood-gas composition of mixed pulmonary venous blood provides information about the effectiveness of pulmonary gas exchange and the overall balance between ventilation and pulmonary capillary perfusion.

Systemic venous blood has a composition that reflects the conditions of the ISF of the tissues from which it drains. Systemic venous blood from all parts of the body mixes in the right ventricle and is ejected into the pulmonary artery as mixed venous blood (Figure 9-1). The blood-gas composition of mixed venous blood provides information on the balance between gas exchange and blood flow in the systemic capillaries. In other words, data obtained from blood-gas analysis of mixed venous blood provide information on the average composition of the ISF of systemic tissues, which, in turn, is the environment of all systemic cells.

The composition of venous blood draining any given systemic tissue depends on the metabolic activity and blood flow in that tissue. For example, if the metabolic rate of a tissue increases, both oxygen uptake and CO_2 production will increase. If the blood flow remains constant or does not increase in proportion to the metabolic rate, the P_{O_2} of the blood perfusing the tissue will decrease and the P_{CO_2} will increase. On the other hand, if blood flow to a tissue increases and the metabolic rate does not change, the P_{O_2} of venous blood draining the tissue will increase because a smaller percentage of the oxygen that is delivered in arterial blood will be used for metabolism. The P_{CO_2} of venous blood leaving the tissue will decrease because more blood flow is available to dilute the added CO_2.

Venous Blood Gases and Assessment of Oxygenation of Tissues

At rest, the effects of different blood flows to organs are reflected in the P_{O_2} of the venous blood leaving those tissues. For example, renal venous blood has a higher P_{O_2} than blood flowing simultaneously in the femoral vein, because renal blood flow is very high compared to the metabolic needs of the kidneys. In fact, because of the rather large contribution of renal venous blood to the total flow in the inferior vena cava (greater than 25%), the P_{O_2} of blood in the inferior vena cava is greater than that of the blood in the superior vena cava. Blood flow in peripheral veins in the extremities will vary depending on the temperature of the limb and muscular activity in the region. Consequently, the P_{O_2} of blood samples obtained from different limb veins will be less uniform in composition because of variabilities in blood flow and, to a certain extent, metabolic rate. Therefore, it has been generally concluded that the P_{O_2} of venous blood varies too much from one site to the next to provide useful data on the P_{O_2} of systemic tissues as a whole.

Blood from all of the systemic veins, however, mixes in the right ventricle and is then ejected into the pulmonary artery. Blood in the pulmonary artery has a composition that represents the flow weighted average from systemic tissues in the body. Therefore, the P_{O_2} and oxygen content of blood in the pulmonary artery represent the "venous average" for the body as a whole. It is not as easy to gain access to the pulmonary artery as it is to sample from peripheral veins. Blood may be sampled from the pulmonary artery through a

Table 9-4 Changes in the Blood-Gas Composition of Arterial and Mixed Venous Blood during Reduction in Cardiac Output*

	Predicted changes in tissues	Arterial blood	Mixed venous blood
pH	⇓	⇓	⇓
P_{CO_2}	⇑	⇓	⇑
HCO_3^-	⇓ †	⇓	⇓ †
P_{O_2}	⇓	⇑ ‡	⇓

*These changes were observed in animal experiments in which the cardiac output was decreased through hemorrhage. See references for Chapter 9 listed at the end of the text.

† The degree of decrease in bicarbonate concentration depends on the severity of the lactic acidosis produced. If the change in P_{CO_2} is greater than the change produced by the lactic acid, the bicarbonate concentration of the ISF might even increase slightly. The bicarbonate concentration in arterial blood will always decrease when the reduced cardiac output is produced by hypovolemia alone.

‡ The P_{O_2} of arterial blood will generally increase somewhat following reduced cardiac output from hypovolemia because of an increase in the ratio of ventilation to perfusion.

flow-directed (Swan-Ganz) catheter. Obviously, placement of this type of catheter should be done only when there is a clear indication of the need for such a device based on the clinical condition of the patient. If such a catheter is in place and there is need to evaluate the status of systemic tissue oxygenation, then the P_{O_2} of mixed venous blood coupled with data on hemoglobin concentration and percent saturation will provide the information needed. Normal values for blood-gas composition of mixed venous blood are shown in Table 9-3.

Venous Blood Gases and Assessment of Acid-Base Status

As stated in Chapter 1, the acid-base status of the body as a whole is best represented by the acid-base status of interstitial fluid. ISF cannot be readily sampled without contamination, but the plasma that comes into equilibrium with it provides the next best fluid for analysis. As was discussed earlier, mixed venous blood-gas composition provides a flow weighted average for all venous blood. In general, as long as cardiac output remains reasonably constant, the pattern of change in blood-gas composition of mixed venous blood will *mimic the pattern of change* seen in arterial blood. In other words, changes in both arterial and mixed venous blood will be in the same direction.

However, when blood flow decreases with decreasing cardiac output, the pattern of change in the mixed venous blood will be different from that seen in arterial blood. These differences become clear if the changes that occur in both arterial and mixed venous blood are compared during reductions in cardiac output (Table 9-4 and Figure 9-12). When cardiac output is decreased,

FIGURE 9-12
The relationship between pH and bicarbonate concentration in mixed venous and arterial blood following a reduction in cardiac output. The blood-gas composition of mixed venous blood (*MV*) falls in the region of combined acidosis whereas that of arterial blood (*A*) falls in the region of compensated metabolic acidosis. The P_{CO_2} isobar at 46 torr is included to show the normal mixed venous P_{CO_2}. *R* is the respiratory compensation curve.

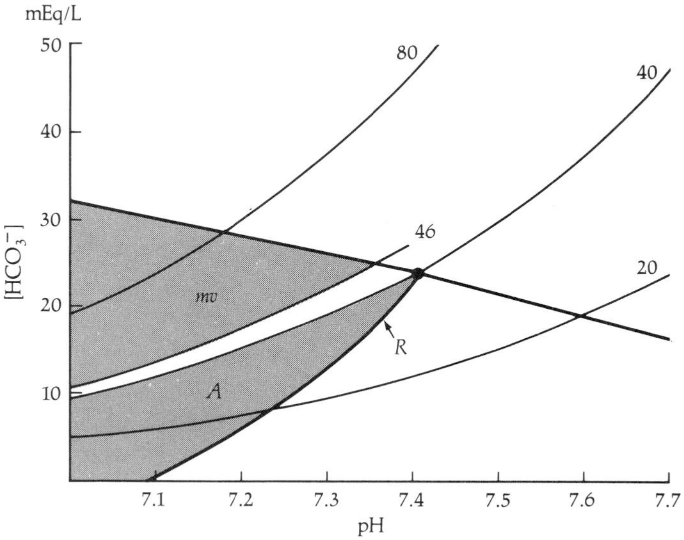

systemic tissue perfusion is reduced. The reduced delivery of oxygen causes tissue hypoxia, and anaerobic metabolism increases. The lactic acid that is produced from this type of metabolism will cause the pH to decrease. Because of the low blood flow, the P_{CO_2} of tissues will also increase. The bicarbonate concentration will usually decrease because of the added lactic acid, but the magnitude of the decrement depends on the severity of the acidosis and the change in P_{CO_2}. If the P_{CO_2} of tissues increases greatly because of reduced flow, bicarbonate concentration of tissues will not decrease as much as expected with this type of metabolic acidosis because increased P_{CO_2} causes bicarbonate concentration to increase. In other words, the increased concentrations of lactic and carbonic acids have opposing effects on bicarbonate concentrations in the tissue ISF (see Equation 5-29). The net result is the production of a combined acidosis in the systemic ISF (Figure 9-12).

The changes that occur in arterial blood (Table 9-4) are consistent with those observed during a compensated metabolic acidosis. The influx of lactic acid into the blood stimulates the peripheral chemoreceptors, causing an increase in alveolar ventilation and a rather quick compensation for the metabolic acidosis. Hence, arterial P_{CO_2} will decrease. Hyperventilation will also occur as a response to hypotension. Thus, both metabolic acidosis and hypotension cause a reflex increase in alveolar ventilation rate. The P_{O_2} of arterial blood will increase slightly because of hyperventilation. Note that the changes in arterial P_{CO_2} and P_{O_2} are opposite those predicted for tissues.

The mixed venous blood-gas composition changes in the same direction as predicted for tissue ISF. The reason that mixed venous blood-gas composition changes in the same direction as that of ISF is because venous blood equilibrates with tissue ISF while the blood is flowing in systemic capillaries. Therefore, it is reasonable to conclude that during reduced cardiac output, mixed venous blood-gas composition should provide a more accurate assessment of both acid-base and oxygenation status of systemic tissues than arterial blood-gas composition. It is also reasonable to conclude that arterial blood-gas composition will never provide a more accurate picture of the status of ISF than mixed venous blood-gas composition.

Peripheral Venous Blood and Acid-Base Assessment

Several investigators have shown that peripheral venous blood may be used in lieu of arterial blood for blood-gas analysis in the assessment of acid-base status. In general, there is excellent correlation between assessment based on arterial and peripheral venous blood. It has been recommended by some that peripheral venous blood be used in lieu of arterial blood when repeated sampling for assessment of acid-base status is needed but pulmonary function is known to be adequate. This would be of particular importance for patients in whom trauma to existing arterial vessels must be minimized, such as patients recovering from diabetic ketoacidosis or in chronic renal failure. It is of interest that no published reports have compared the blood-gas composition of peripheral venous blood to that of mixed venous blood.

Admonitions on the Technique of Sampling Mixed Venous Blood

Certain techniques must be observed when obtaining samples of mixed venous blood from flow-directed catheters. First, the balloon at the end of the catheter must be deflated. Second, the sample must be drawn slowly, usually over a period of about 60 seconds. The reason for such slow withdrawal is that the catheter is usually advanced downstream from the main trunk of the pulmonary artery into a branch of the pulmonary vasculature where it may be wedged to measure downstream pressures (wedge pressures). In this position, if a sample is drawn too quickly, blood will be aspirated from the pulmonary capillaries where exchange with alveolar gas has already occurred. Such a "contaminated" sample would have a P_{O_2} greater than that of mixed venous blood. Therefore, the sample must be drawn slowly to avoid contamination. Third, the catheter is usually long and contains about 1 ml of dead space, usually filled with a sterile solution of 5% glucose or saline. This volume must first be removed and discarded, but again the rate of removal must be slow to avoid drawing pulmonary capillary blood into the catheter. Finally, care must be taken to avoid clotting of blood within the catheter. As soon as the sample is obtained, the catheter must be flushed with sterile saline or glucose solution. There can be no delay in flushing because the slow rate of removal and catheter length increase the probability of blood clotting within the lumen of the catheter.

Clinical Examples

CHAPTER

10

The case examples described in this chapter were obtained from several different acute care settings and are presented here to illustrate different kinds of fluid, electrolyte, and acid-base disturbances. In each case a short history with laboratory data is presented and then interpreted. Frequently serial sets of data are presented to show a clearer picture of changes that occurred. Values for P_{CO_2} and P_{O_2} in the tables of serial sets of blood gases are given in units of torr and those for $[HCO_3^-]$ are in units of milliequivalents per liter. These case examples are only a small sample of pathophysiologic processes that can cause fluid, electrolyte, and acid-base disturbances.

In each case example, try to assess the fluid, electrolyte, and acid-base status of the patient from the data presented. Then read the interpretations of the data and compare your assessment with the one provided. Before proceeding into these cases the reader who is unfamiliar or only vaguely familiar with the concept of anion gap should read Appendix 6.

CASE 1

A 54-year-old man was admitted to a large metropolitan hospital for surgical repair of an abdominal aortic aneurysm. After surgery, the patient was brought to the surgical intensive care unit. An endotracheal tube, nasogastric tube, and urinary catheter were in place. He was mechanically ventilated with an inspired gas mixture containing 40% O_2. His tidal volume was set at 950 ml and he was breathing at a frequency of 12 breaths per minute. He was receiving a diuretic (Lasix®) twice a day and morphine for pain. Fluid therapy consisted of 2 L of 5% glucose in water (D5W) every 24 hours. Forty-eight hours after surgery the results of arterial blood-gas analysis were pH 7.62; P_{CO_2}, 26 torr; [HCO_3^-], 27 mEq/L; and P_{O_2}, 74 torr. Serum electrolyte concentrations were [Na^+], 126 mEq/L; [K^+], 3.4 mEq/L; [Cl^-], 83 mEq/L; and [Ca^{++}], 4.8 mEq/L.

This is the case described in Chapter 1. The patient developed a rather severe combined alkalosis following surgery (region IV in Figure 6-4). The respiratory component was due to hyperventilation caused by the high minute volume set by mechanical ventilation. The metabolic component was caused by two processes. First, gastric secretions were lost through nasogastric suction without adequate replacement of chloride and volume. As discussed in Chapter 6, this loss leads to increased generation of bicarbonate by the parietal cells of the gastric mucosa. Second, Lasix® is a powerful diuretic that blocks Cl^- reabsorption in the ascending limb of Henle's loop. The diuretic causes a pronounced diuresis and volume contraction. The resulting decreased blood pressure activates the renin-angiotensin-aldosterone mechanism that stimulates conservation of filtered Na^+. Because of the chloride deficit and the increased delivery of Na^+ to the distal nephron caused by the action of the diuretic (Chapter 4), there is an increased secretion of both H^+ and K^+ in exchange for Na^+ in the distal nephron. The net result is that bicarbonate is generated to accompany sodium although the plasma bicarbonate concentration and pH are both increased.

The hyponatremia was caused by the renal loss of Na^+ from the action of diuretics and from loss of Na^+ in gastric secretions removed by nasogastric suction. The sodium was not replaced, but water was partially replaced. Hypochloremia was induced by loss of Cl^- in gastric secretions and again was not replaced. The hypokalemia was caused by the action of the diuretic (see Figure 4-12) and some was also lost with removal of gastric secretions. Note that the anion gap (Appendix 6) was increased. This was likely at least partly due to increased lactate production that accompanies the hypocapnia in the respiratory component of the alkalosis.

The respiratory component of this acid-base disturbance could have been corrected by increasing the P_{CO_2}. Adding dead space to the ventilator will increase P_{CO_2}. There is certainly a concern that decreasing alveolar ventilation might decrease P_{O_2} to dangerously low levels. Note that the $F_{I_{O_2}}$ was 0.4. However, one must question how effective O_2 delivery is at a pH of 7.62 and markedly decreased plasma volume. The alkalosis shifts the oxyhemoglobin dissociation curve to the left (Bohr effect), hence, tissue P_{O_2} must be decreased to receive the same amount of O_2 during alkalosis as is obtained at normal values of pH. Adding to the problem is a decreased cardiac output from volume contraction, which causes O_2 delivery to be seriously compromised.

Correction of the metabolic component requires replacement of Cl^- lost from drainage of gastric fluid and some of the Cl^- lost from diuretic therapy. Provided the patient could tolerate a fluid expansion, normal saline with added K^+ is the therapy of choice because it provides Cl^- (154 mEq/L) and also expands the ECF, thereby diluting HCO_3^- in plasma. The K^+ is needed because the patient is K^+ depleted from alkalosis and diuretics. Fluid volume expansion also acts as a negative feedback to decrease renin secretion (see Figure 2-7).

CASE 2

A 37-year-old female was admitted to the medical service complaining of nausea, vomiting, shortness of breath, productive cough, and pain in the right upper quadrant of her abdomen and left chest wall. She stated that she had not eaten in the last three days and had eaten very little in the past three weeks. She was a chain smoker and had smoked two to three packs of cigarettes per day for 18 years. She also had active peptic ulcer disease. Her vital signs were temperature, 36.5°C; respiratory rate, 48 per minute; blood pressure, 100/48 mmHg, and heart rate at rest 120/minute. She weighed 42 kg (92 lb) and had decreased skin turgor at the time of admission to the hospital.

Arterial blood gases were pH, 7.49; P_{CO_2}, 29 torr; [HCO_3^-], 22 mEq/L; and P_{O_2}, 57 torr. Serum electrolyte concentrations were [Na^+], 132 mEq/L; [K^+], 3.0 mEq/L; [Ca^{++}], 7.4 mg/dl; and [Cl^-], 92 mg/dl. Her BUN was 26 mg/dl and serum creatinine concentration was 1.2 mg/dl. Total serum protein was 4.8 g/dl and albumin was 2.5 g/dl. Normal values vary between 10 and 20 g/dl for BUN, 6 and 8 g/dl for total protein, and 3.5 and 5 g/dl for albumin. Her hemoglobin concentration was 10.7 g/dl.

This patient had both pneumonia and gallbladder disease. Her pain was caused by inflammation of the gallbladder and pleural wall. Bleeding from her chronic ulcer disease and obvious malnutrition were causing her anemia and weight loss.

She developed a respiratory alkalosis that was not compensated. This can be seen by the fact that [HCO_3^-] falls on the CO_2 buffer line. The immediate explanation might be that the hyperventilation had been present for too short a period of time to permit compensation. This explanation is unlikely considering the evidence of long-standing disease. Compensation would have required increased renal excretion of HCO_3^-, which clearly was not occurring. There are several opposing factors that, taken as a whole, explain the lack of compensation for the alkalosis. Some of these factors tend to decrease [HCO_3^-]—hypoxia, lactic acid production from hypocapnia, and decreased arterial P_{O_2}. The hypoxia and hypocapnia cause increased lactic acid production, which is buffered with HCO_3^-. Two kinds of hypoxia were present, arterial hypoxemia (P_{O_2} only 57 torr) and ischemia from hypovolemia. The decreased P_{CO_2} itself also tends to decrease [HCO_3^-] through the equilibrium shown in Equation 5-31.

Factors that tended to increase [HCO_3^-] included increased HCO_3^- generation by the gastric mucosa, volume contraction itself, and increased generation of HCO_3^- by the kidneys. People with peptic ulcer disease have increased rates of secretion of HCl by the gastric mucosa. This HCl was being lost through vomiting. The net result is that for every H^+ secreted into the gastric lumen, a HCO_3^- is added to the blood (see Figure 6-3). Thus, some HCO_3^- was being generated by the parietal cells of the stomach. Volume contraction itself tends to concentrate HCO_3^-. The renal tubules were generating increased amounts of HCO_3^- to accompany the sodium retention because of the volume depletion. Recall that volume depletion activates the renin-angiotensin-aldosterone mechanism, as discussed in Chapters 2, 4, and 7. The net result of those opposing processes left the [HCO_3^-] at 22 mEq/L. This example points out the complexity of the factors involved in regulating [HCO_3^-].

The hyponatremia was accompanied by volume contraction. Evidence for volume contraction in this case includes reduced blood pressure, decreased skin turgor, and increased BUN with a normal creatinine concentration. The hyponatremia was probably caused by decreased sodium intake (she had not eaten for at least three days) and loss of sodium with emesis and urine. Also, ADH secretion was probably increased causing retention of free water, which diluted [Na^+] in ECF.

The hypokalemia is a concomitant of alkalosis and loss of gastric secretions. Hypocalcemia accompanied the decreased plasma protein concentrations. Although she had alkalemia, the patient did not develop physical signs of hypocalcemia because protein concentration was also diminished and her tissues were probably not as alkalotic as indicated by arterial pH. Recall that decreased cardiac output causes retention of CO_2 in tissues and increased lactic acid production. The fact that her anion gap (Appendix 6) was increased to 21 mEq/L supports the contention that serum lactate concentration was probably increased.

CASE 3

A 46-year-old man was hospitalized because of nausea, jaundice, and severe abdominal pain. On physical examination he was pale, diaphoretic, and his abdomen was extremely tender, making palpation nearly impossible. A Computerized Axial Tomography (CAT) scan was performed, and a tentative diagnosis of a subdiaphragmatic abscess was made. His vital signs prior to surgery were blood pressure, 140/100 mmHg; pulse, 124/minute; respiratory rate, 28/minute; and temperature, 39°C.

An exploratory laparotomy was performed and a completely necrotic pancreas was found. A total pancreatectomy was performed. During surgery the patient lost an estimated 19 L of blood. The volume replaced included 26 U of whole blood, 8 L of normal saline, 4 L of lactated Ringer's, 1 L D5 in 0.45 normal saline with dopamine, and 7 U of fresh frozen plasma. The surgical procedure lasted 9 hours. In addition to the total pancreatectomy, a splenectomy, gastrojejunostomy, and cholecystojejunostomy were performed.

Table 10-1 Arterial Blood-Gas Composition for Case 3

Set	pH	P_{CO_2} torr	$[HCO_3^-]$ mEq / L	P_{O_2} torr
1	7.15	41	14	75
2	7.29	40	19	71
3	7.40	29	18	102

After surgery, the patient was admitted to the intensive care unit and ventilated mechanically with a tidal volume of 800 ml at 12 per minute with assist. His $F_{I_{O_2}}$ was 0.6. His first set of postoperative arterial blood gases is shown in Table 10-1. The first set of gases shows that the patient was in an uncompensated metabolic acidosis. This severe acidosis was due to circulatory shock. The acidosis was uncompensated because of the controlled ventilation. In addition to the circulatory shock, CNS depression might have existed because of anesthetic agents and narcotics. He was given two ampules of sodium bicarbonate (100 mEq) and a second set of blood gases was obtained 30 minutes later. The patient was still in an uncompensated metabolic acidosis. At this time the tidal volume was increased to 1000 ml and the rate set at 14 per minute. One hour later the patient's blood-gas composition stabilized, as shown in set 3. For the next 3 hours this patient had the kind of mixed acid-base disorder exemplified by set 3, namely, a metabolic acidosis with a superimposed respiratory alkalosis (region VI, Figure 6-4). This type of mixed disturbance may be seen when mechanical hyperventilation is superimposed with a metabolic acidosis. This patient died about 4 hours after set 3 was obtained.

CASE 4

A 49-year-old man with chronic bronchitis was admitted to the hospital because he had increasing dyspnea for five days. On admission, his temperature was 38°C; pulse, 90/minute; blood pressure, 115/70 torr; and respiratory rate, 30/minute. The patient had pitting edema in both ankles, rales in the bases of both lungs, and many premature ventricular contractions (PVCs), as shown on his electrocardiogram. Table 10-2 shows the results of arterial blood-gas analysis. On admission, the patient had a compensated respiratory acidosis (see Table 6-3). He was also diagnosed as having heart failure. Because of his deteriorating condition he was intubated and mechanically ventilated. He was given diuretics and digitalis for his heart failure. Six hours after admission, the second set of blood gases was obtained.

This patient had a chronic and compensated respiratory acidosis because of chronic lung disease. However, heart failure complicated the situation and his initial pH was lower than usual for him (set 1). The decrease in pH was due to

Table 10-2 Arterial Blood-Gas Composition for Case 4

Set	pH	P_{CO_2} torr	$[HCO_3^-]$ mEq/L	P_{O_2} torr
1	7.28	66	30	49
2	7.33	58	30	55

Table 10-3 Arterial Blood-Gas Composition for Case 5

Set	pH	P_{CO_2} torr	$[HCO_3^-]$ mEq/L	P_{O_2} torr
1	7.06	86	24	44
2	7.33	43	22	97

inadequate circulation of blood, which usually increases lactic acid production. The increased circulation following therapy helped improve his acid-base status (set 2). When he was discharged, he was sent home on medications, which included diuretics, potassium supplements, and O_2 at 2 L/minute.

CASE 5

A 19-year-old female was taken to the emergency room after being thrown from her horse. She was unconscious on admission and exhibited an irregular breathing pattern. Her arterial blood gases are shown in Table 10-3, set 1. Because of these results, she was intubated and mechanically ventilated. Twenty minutes after beginning mechanical ventilation her blood-gas composition had changed to that shown in set 2.

This patient developed a combined acidosis from the head injury she had sustained in her fall from the horse. The respiratory component was due to hypoventilation and the consequent retention of CO_2. The metabolic component was due to hypoxemia, which caused an increase in anaerobic metabolism and increased production of lactic acid. This case shows that metabolic acidosis can be caused by respiratory dysfunction. Restoring ventilation to a reasonably normal value reduced P_{CO_2} and improved the P_{O_2}, hence, correcting the acid-base imbalance.

CASE 6

A 40-year-old woman was admitted to a community hospital with abdominal pain, jaundice, and nausea. She stated that the epigastric pain and nausea had

Table 10-4 Arterial Blood-Gas Composition for Case 7

Set	pH	P_{CO_2} torr	$[HCO_3^-]$ mEq / L	P_{O_2} torr
1	7.51	26	20	71
2	7.49	24	18	70
3	7.26	51	22	84

been present for two to three weeks. She smoked $1\frac{1}{2}$ packs of cigarettes and drank three to four hard liquor drinks a day. On physical examination a firm, tender mass 7 cm in diameter and 7 cm below the costal margin was palpated. Laboratory tests and roentgenograms confirmed obstructive jaundice. At surgery a pancreatic pseudocyst, 8 cm in diameter, was excised. The patient had an uneventful postoperative course until four days after surgery. She developed a spiking fever of 39°C, bowel sounds became faint, and there were decreased breath sounds in the base of the right lung. Arterial blood-gas analysis revealed pH, 7.48; P_{CO_2}, 30 torr; and $[HCO_3^-]$, 22 mEq/L. The P_{O_2} was 66 torr.

This patient developed a postoperative pulmonary infection. The fever and developing pneumonia caused her to hyperventilate and develop an uncompensated respiratory alkalosis. Insufficient time had lapsed to allow for a significant degree of compensation. Antibiotic therapy was begun, and the patient had an uneventful recovery.

CASE 7

A 49-year-old man developed a fever after coronary bypass surgery. He was no longer being mechanically ventilated and until this point had progressed satisfactorily. Because of his deteriorating condition and the possible presence of a postoperative infection, he was put in isolation in the surgical intensive care unit. His arterial blood gases are shown in Table 10-4.

In set 1, a respiratory alkalosis was present that was beginning to be compensated. Evidently, this man had been hyperventilating for several hours. Set 2 was taken 12 hours after antibiotic therapy was begun. There was no improvement in the alkalosis, but compensation had developed. Two hours after set 2 was obtained, the patient had a respiratory arrest. He was intubated and mechanically ventilated. Set 3 was obtained 25 minutes after mechanical ventilation was begun and showed that the patient had a combined acidosis (Figure 10-1). The respiratory component was due to the increased P_{CO_2}. The metabolic component was present before respiratory arrest. Because the decrement in $[HCO_3^-]$ could not be made up quickly, the combined acidosis developed when hypercapnia developed. Essentially the body fluids titrated up the CO_2 buffer line (point 2 to point 3, Figure 10-1) by addition of CO_2.

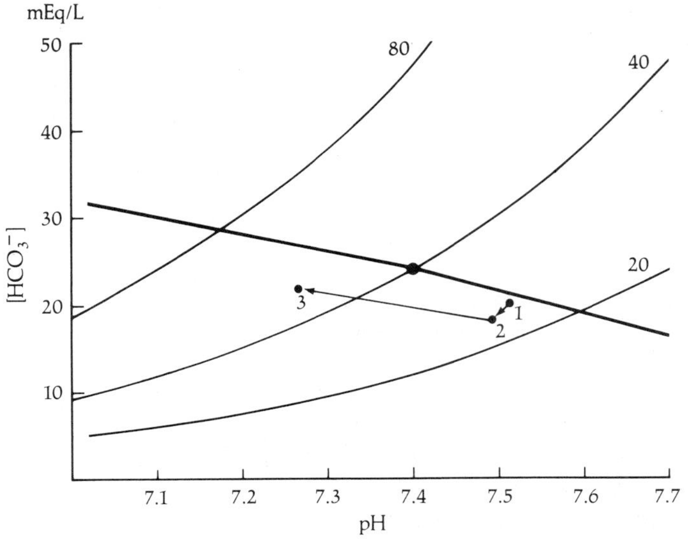

FIGURE 10-1
A pH / $[HCO_3^-]$ graph of the blood-gas data from Table 10-4. Points 1 and 2 are within the area of compensated respiratory alkalosis. Point 3 lies within the region of combined acidosis.

This case illustrates the usefulness of knowing the previous acid-base status of the patient. If point 3 were the only blood-gas data available, the metabolic component would be difficult to explain. In fact, one would be tempted at first to explain the metabolic component as being due to lactic acid production from hypoxemia during the respiratory arrest. However, when $[HCO_3^-]$ did not return to normal within a few hours, another source for the metabolic component would have to be sought. It is clear in this case that the previous compensation is the source.

This patient died a few days later from gram-negative septicemia. The initial alkalosis was most likely caused by the sepsis and fever. The precise mechanism by which sepsis induces hypocapnia is not known, but a direct stimulus by bacterial toxins on chemoreceptors has been postulated.

CASE 8

A 65-year-old comatose female was brought to the emergency room at 7:15 A.M. On admission, her vital signs were pulse, 88/minute; respiratory rate, 14/minute; blood pressure, 110/54 mmHg; and temperature, 36°C (rectal). She also had poor skin turgor and a strong smell of acetone on her breath. Tables 10-5 and 10-6 show the results of laboratory tests.

When the patient came to the emergency room she was in a diabetic coma and had a severe metabolic acidosis, which was compensated (set 1, Table 10-5). Initial treatment consisted of giving insulin and beginning an infusion of 1 L of normal saline and intravenous infusion of 80 mEq of $NaHCO_3$. Fifteen minutes after the $NaHCO_3$ was given, a second set of blood gases was

Table 10-5 Arterial Blood-Gas Composition for Case 8

Time	pH	P_{CO_2} torr	$[HCO_3^-]$ mEq / L	P_{O_2} torr
7:30 A.M.	7.05	19	5.1	72
8:30 A.M.	7.19	22	8.1	84
2:20 P.M.	7.30	41	19.5	116

Table 10-6 Serum Electrolyte Concentrations for Case 8

Time	$[Na^+]$ mEq / L	$[K^+]$ mEq / L	$[Cl^-]$ mEq / L	[Glucose] mg / dl	[BUN] mg / dl	[Creatinine] mg / dl
7:45 A.M.	134	5.8	100	958	24	2.2
9:40 A.M.	148	3.2	110	675	—	—
3:30 P.M.	147	3.6	109	348	—	—

obtained. At 8:30 A.M. (set 2) it can be seen that the patient had been titrated to a pH of 7.19, with little change in P_{CO_2}. Another 40 mEq of $NaHCO_3$ was given at 9:50 A.M.

The metabolic acidosis was produced from excess production of ketoacids (acetoacetic and β-OH-butyric acids), which are generated from lipid catabolism. Normally, insulin helps regulate fat catabolism in adipose tissue and liver. When insulin is not available, fat breakdown proceeds in an uncontrolled fashion, with resultant production of excess ketoacids. The acids are then buffered, as shown in Figure 5-9*A* and Equation 6-1, in which *HB* represents addition of ketoacids. Giving insulin not only helps control glucose concentration, but also decreases the rate of lipolysis and ketoacid production.

The acidosis was quite severe in this patient, and HCO_3^- was needed to bring pH into a safe range. The initial 80 mEq of HCO_3^- titrated the pH from 7.05 to 7.19 along the 20-torr isobar (Figure 10-2). Little change in P_{CO_2} would be expected in this short a period of time because the added HCO_3^- had not equilibrated across the blood-brain barrier. When pH reached 7.3 (Table 10-5), administration of more HCO_3^- was unnecessary because the patient could then correct pH to normal through her own renal and respiratory function. Intravenous infusion of HCO_3^- to correct the $[HCO_3^-]$ from 5.1 to 24 mEq/L, although technically possible, was contraindicated because time is needed for HCO_3^- in blood to equilibrate with CSF, as discussed in Chapter 4. If enough HCO_3^- had been given initially to correct her bicarbonate deficit in the ECF, she would have developed a combined alkalosis and her pH would have been greater than 7.6 (point 4, Figure 10-2).

The results of chemical analysis of serum are shown in Table 10-6. Initially (7:45 A.M.), she was hyperkalemic, hyperglycemic, and dehydrated to the point that she was in prerenal failure, as shown by the increased BUN and creatinine

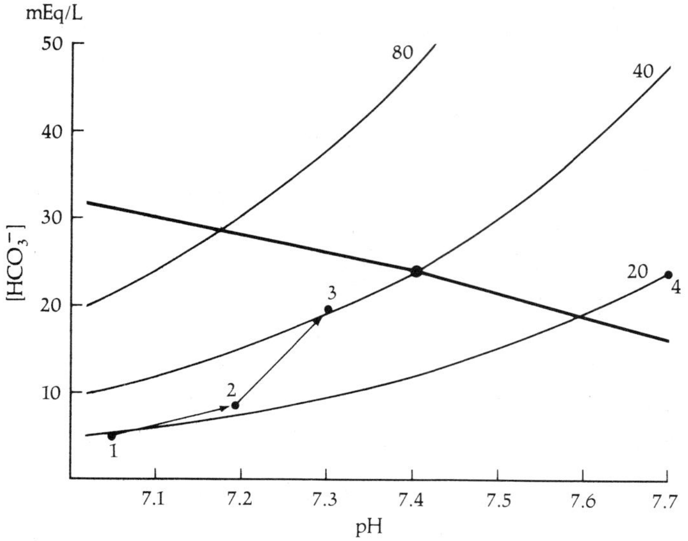

FIGURE 10-2
A pH / [HCO_3^-] graph of the data from Table 10-5. This is a classic example of compensated metabolic acidosis. Point 4 shows the effects of what would have happened if the HCO_3^- deficit had been completely corrected on admission.

concentration. This is an example of severe hyperosmotic volume contraction. Hence, administration of fluids was of critical importance in this patient. At 9:40 A.M. a second analysis of serum electrolyte concentrations revealed a markedly decreased K^+ concentration. As a result, a second liter of saline containing 40 mEq of K^+ was given. It is important to recognize that the patient had a K^+ deficit when she was brought to the hospital although she was hyperkalemic. Recall that K^+ enters the ECF in exchange for H^+ across cell membranes when fixed acids are added to body fluids (see Figure 5-9*A*). The kidneys, if functioning, will excrete the K^+ that was exchanged. Furthermore, extra K^+ was excreted in this case because of increased exchange of Na^+ for K^+ in the distal nephron (see Chapter 4). There was increased delivery of Na^+ to the distal nephron because sodium is the cation that accompanies the conjugate bases of the ketoacids.

The patient was debilitated and unable to replace her fluid losses and she became dehydrated. Evidence for dehydration was present in the physical findings (low blood pressure and decreased skin turgor) and in the laboratory data (azotemia). The patient developed oliguria and prerenal failure while K^+ was still exchanging for H^+ across cell membranes. The result was hyperkalemia although there was a total deficit of K^+ in the body. The combination of insulin, volume expansion, and $NaHCO_3$ reversed K^+–H^+ exchange and serum [K^+] decreased to 3.2.

The initial [Na^+] shown in Table 10-6 indicates hyponatremia. It must be remembered that this patient had hyperosmotic volume contraction from hyperglycemia with subsequent osmotic diuresis (see Figure 8-1). The final effect on [Na^+] depends on the amount of dilution caused by fluid shift from the ICF versus renal excretion of water with subsequent diuresis. Furthermore, this patient lost volume through insensible routes that was not replaced because she was comatose. On admission to the emergency room she had a mild hyponatremia but a marked deficit of total sodium in her ECF because of volume depletion. Sodium concentration in the serum was increased at 9:40 A.M. and 3:30 P.M. because of volume replacement with 2 L of normal saline (154 mEq/L) and 120 mEq of $NaHCO_3$. Added to this was the shift of water back into the ICF accompanying glucose uptake by cells following insulin therapy.

CASE 9

A 54-year-old businessman who was a buyer for a large chain of department stores was admitted to the medicine floor because of persistent abdominal pain, nausea, and weight loss. He had been vomiting for three days and his last emesis had been dark brown in color. He had smoked one pack of cigarettes per day for 30 years and had a chronic cough. On physical examination the patient was found to be pale, diaphoretic, with a tender abdomen over the epigastrium. He had been taking antacids to help control his abdominal pain. His vital signs on admission were temperature, 37°C; pulse, 110/minute; respiratory rate, 14/minute; and blood pressure, 100/66 mmHg. Arterial blood gases were pH, 7.49; P_{CO_2}, 47 torr; [HCO_3^-], 36 mEq/L; and P_{O_2}, 74 torr.

The blood gases indicate the patient had a compensated metabolic alkalosis. The alkalosis was caused by volume depletion and generation of excess HCO_3^- by the gastric mucosa. The loss of gastric fluid causes loss of volume and Cl^- and generation of HCO_3^-, which is added to the blood (see Figure 6-3). The volume depletion stimulates aldosterone production by means of the renin-angiotensin mechanism. The kidneys are then stimulated to conserve Na^+ and H_2O to minimize volume contraction. Because there was also a Cl^- deficit with loss of gastric fluid, the kidneys had to generate more H^+ to exchange for Na^+. Consequently, renal HCO_3^- generation also increased and Na^+ was reabsorbed with HCO_3^-. This patient had alkalosis long enough for HCO_3^- to cross the blood-brain barrier, and hence he was hypoventilating in compensation for the alkalosis.

CASE 10

A 29-year-old male was admitted to a community hospital following two days of severe abdominal pain, nausea, and vomiting. On physical examination he was found to have a large mass with tenderness in the epigastrium and diminished bowel sounds. The following data were obtained from analysis of his serum: Amylase was 1750 U (normal is 20–85 U); triglycerides were 4300 mg/dl (normal is 20–160 mg/dl); glucose was 475 mg/dl; total serum protein was 7.6 g/dl; [Ca^{++}] was 6.9 mg/dl; [Na^+] was 128 mEq/L; [Cl^-] was 101 mEq/L; and [K^+] was 4.2 mEq/L. His hematocrit was increased to 51% of total blood volume and hemoglobin concentration was 20.6 g/dl. His serum was milky and opaque to light; in fact, the laboratory technician described the serum as "cream."

This young man was diagnosed as having acute pancreatitis from hyperlipidemia. His hyponatremia was caused by two factors. First, glucose concentration in his serum was 380 mg/dl greater than normal. (Pancreatitis frequently causes a diminished secretion of insulin.) As described in Appendix 5, Na^+ concentration in serum will decrease approximately 1.5 mEq/L for every 100 mg/dl increment in glucose concentration. Thus, the hyperglycemia should cause [Na^+] to decrease by 5 to 6 mEq/L. Therefore, hyperglycemia should cause a decrease in serum [Na^+] to about 134 to 135 mEq/L. The patient's severe hyperlipidemia caused a further decrease in serum sodium concentration because of pseudohyponatremia. If it is assumed that the lipids and proteins in his serum together occupy 11.9 ml of every 100 ml of serum, that leaves 88.1 ml of plasma water in which the sodium should be distributed. Assuming that serum Na^+ concentration corrected for the increased glucose concentration should be 135 mEq/L, every liter of plasma water should contain 146 mEq/L based on this patient's total protein concentration of 7.6 g/dl. This result is calculated as follows: The sodium is dissolved in the water phase of the plasma. Assuming that 7.6 ml of every deciliter is protein, 92.4 ml of every deciliter remains as water for dissolving Na^+. Hence, Na^+ concentration in plasma water was 135 mEq/0.924 L plasma water = 146 mEq/L. In this patient, however, each liter of plasma contained only 881 ml of plasma water because protein plus lipid occupies about 11.9% of the total volume, that is, 119 ml. Therefore, this patient has only 88.1% of the total 146 mEq Na^+ in each liter of plasma. Multiplying,

$$146 \times \frac{88.1}{100} = 129 \text{ mEq / L of serum or plasma} \qquad (10\text{-}1)$$

which is essentially his serum sodium concentration. The patient, therefore, has a true decrease in total sodium concentration in his plasma water plus pseudohyponatremia caused by hyperlipidemia.

The patient also has hypocalcemia, probably caused by precipitation in the fluid accumulating in a pseudocyst in and around the pancreas.* There is evidence of third space accumulation of fluid in the pseudocyst and peritoneal cavity, as shown by the increase in hematocrit and hemoglobin concentration of his blood.

---

*The hyperlipidemia accounts for only 0.3 mg/dl of the deficit in [Ca^{++}].

Table 10-7 Arterial Blood-Gas Composition for Case 11

Set	pH	P_{CO_2} torr	$[HCO_3^-]$ mEq/L	P_{O_2} torr
1	7.53	18	15	122
2	7.24	30	12	106
3	7.13	18	6	89
4	7.22	36	14	72
5	7.47	29	21	78
6	7.53	32	27	80
7	7.59	32	31	86

CASE 11

A 55-year-old man in good health was a personnel manager for a large company. He was working in his office when he experienced "crushing" substernal pain. Twenty minutes later he was admitted to the emergency room for an apparent myocardial infarction. He was extremely anxious and complained of shortness of breath. He was given O_2 by mask at 4 L/minute, and a chest x-ray and electrocardiogram were taken. He was given morphine for pain and Lasix® intravenously. The first of several samples of arterial blood was also taken for blood-gas analysis. The results are shown in Table 10-7. The chest x-ray showed increased vascular markings and a few crepitant rales were heard in the bases of the lungs.

In set 1 it appears that the patient had a compensated respiratory alkalosis. This is the region that the blood-gas composition fits into in Figure 6-4, but it is too soon for compensation to have occurred. Thus, the reduced HCO_3^- must be due to production or addition of fixed acid or a previously existing HCO_3^- deficit. Under the circumstances, increased production of fixed acid is the more likely cause. The patient's new infarction had compromised his circulation and the $[HCO_3^-]$ was probably decreased from lactic acid production. Although his arterial P_{O_2} was high (122 torr), lactic acid was produced because of reduced delivery of O_2 to systemic tissues (ischemic hypoxia).

The patient was admitted to the coronary care unit with a diagnosis of an acute anterior myocardial infarction. Forty minutes later, a second set of blood gases was obtained, revealing the suspected metabolic acidosis that was compensated (Figure 10-3). Two hours later he had a cardiac arrest and the electrocardiogram showed ventricular fibrillation. During resuscitation he was given 100 mEq of HCO_3^-. The third set of blood gases was taken right after resuscitation. Because of the severe metabolic acidosis, another 50 mEq of $NaHCO_3$ were given. Thirty minutes later a fourth set of blood gases showed that the patient was in a compensated metabolic acidosis (Figure 10-3).

During the time that cardiac output was reduced because of the weakened myocardium, lactic acid accumulated in tissues because of ischemic hypoxia of

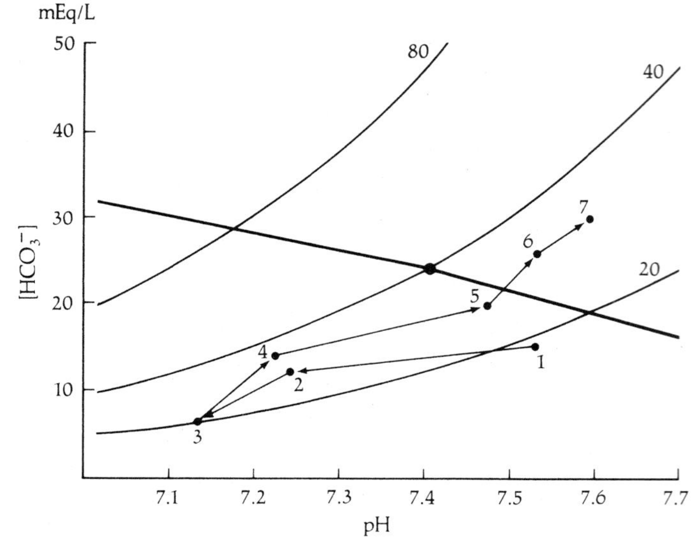

FIGURE 10-3
A pH / $[HCO_3^-]$ graph of the blood-gas data in Table 10-7. Point 1 was obtained shortly after the patient arrived at the emergency room.

tissues. As the patient's cardiac output improved with treatment and bedrest his arterial pH increased. In sets 5, 6, and 7 arterial pH and $[HCO_3^-]$ increased for several reasons. First, improved cardiac output increases O_2 delivery to tissues and reduces anaerobic metabolism, thereby decreasing lactic acid production. The improved oxygenation of tissues, especially liver, allows lactic acid to be metabolized with the concomitant production of HCO_3^- (see Figure 7-5). Second, the reduced cardiac output causes stimulation of Na^+ reabsorption because of increased renin secretion by the kidneys. Because patients with myocardial infarctions are usually treated with diuretics, the decreased volume in the ECF also stimulated renin secretion and concentrated the HCO_3^- remaining in the ECF. The effect of this volume contraction on acid-base status is shown in Figure 10-3 and Table 10-7. The high arterial pH (points 6 and 7 in Figure 10-3) may be detrimental to oxygenation of the myocardium because of the decreased P_{50} with alkalemia. In this particular case, the patient recovered.

CASE 12

A 52-year-old female was brought to the emergency room because of sudden onset of viral symptoms. This woman had been seen several times in the emergency room because of her chronic obstructive pulmonary disease (COPD). She had smoked $2\frac{1}{2}$ packs of cigarettes per day for 36 years. On this occasion she complained of shortness of breath, headache, nausea, vomiting, and diarrhea. Her vital signs were temperature, 39°C; pulse, 64/minute; respiratory rate, 24/minute; and blood pressure, 111/66 mmHg. A chest x-ray

Table 10-8 Arterial Blood-Gas Composition for Case 12

Set	pH	P_{CO_2} torr	$[HCO_3^-]$ mEq / L	P_{O_2} torr
1	7.26	68	30	50
2	7.34	58	30	62
3	7.38	53	30	77
4	7.52	50	40	110
5	7.59	41	38	126
6	7.54	49	41	74
7	7.49	55	41	88
8	7.42	60	38	69
9	7.41	56	34	66
10	7.38	62	36	62
11	7.39	65	40	59

showed right lower lobe infiltrates, cor pulmonale, and congestive heart failure. The results of blood-gas analysis are shown in Table 10-8.

The first set blood gases show that the patient was in a compensated respiratory acidosis. The diagnosis was severe respiratory distress secondary to pneumonia, COPD, cor pulmonale, and left heart failure.

She was transferred to the medical intensive care unit, intubated, and mechanically ventilated with an $F_{I_{O_2}}$ of 0.28 and an intermittent mandatory volume (IMV) of 12. She was also given antibiotics and aminophyllin. A nasogastric tube was inserted because of persistent vomiting of large volumes of guaiac positive emesis. She was given antacids and a maintenance dose of Lasix®. At that time her serum electrolyte concentrations were $[Na^+]$, 128 mEq/L; $[Cl^-]$, 89 mEq/L; and $[K^+]$, 4.1 mEq/L. A second set of arterial blood gases was obtained 40 minutes after mechanical ventilation was begun. The pH and P_{O_2} both increased whereas P_{CO_2} was reduced. Sets 3, 4, and 5 were obtained over the next 24 hours. During that time the increase in pH, decrease in P_{CO_2}, and a marked increase in P_{O_2} continued. There was a shift from compensated respiratory acidosis to uncompensated metabolic alkalosis (Figure 10-4).

Because of the improvement in her respiratory function, the process of weaning the patient from the respirator was begun. The results of this process are shown in sets 6 through 11 in Table 10-8 and Figure 10-4. During the weaning process the patient's acid-base status shifted from uncompensated metabolic alkalosis to mixed respiratory acidosis and metabolic alkalosis. The respiratory component is from her COPD. The metabolic component is maintained by volume contraction of her extracellular fluid from diuretic therapy and nasogastric suction. The diuretics were used to treat her congestive heart failure.

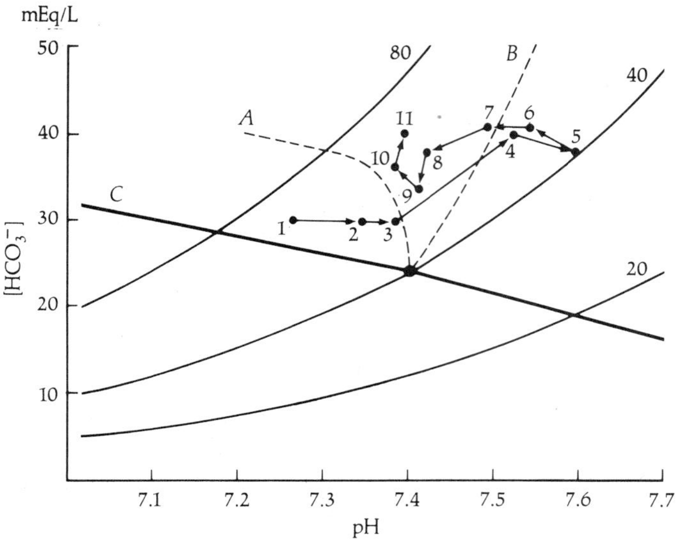

FIGURE 10-4
A pH / $[HCO_3^-]$ graph of the data from Table 10-8. (*A*) The renal compensation curve, (*B*) the respiratory compensation curve, and (*C*) the CO_2 buffer curve. The space between *A* and *B* is the area of mixed respiratory acidosis and metabolic alkalosis.

This mixed type of acid-base disturbance is not as common as the compensated or combined forms. Nonetheless, it is seen occasionally and it must be remembered that this kind of patient has difficulty maintaining pH within safe limits. If there is a change in either component, marked and even life-threatening shifts in pH may occur.

There were several important features of her volume and electrolyte status. First, she had hyponatremia on admission. Second, her blood pressure and tissue turgor were reasonably normal. Third, when she was given sodium, she excreted the sodium in the urine. Fourth, her urine osmolality was 490 mOsm/kg water whereas her plasma osmolality was decreased from a normal of 290 to 265 mOsm/kg water. Fifth, she had a normal serum BUN and creatinine concentration. Finally, she had no evidence of adrenal dysfunction. All six findings are consistent with the diagnosis of SIADH (Chapter 7). This syndrome is sometimes seen in patients with pneumonia. With water restriction, the $[Na^+]$ in her serum returned to normal.

CASE 13

A 22-year-old female was transferred to the medical intensive care unit from the obstetrics floor following the delivery of a normal healthy child. She had developed a cardiac myopathy of unknown origin and was in cardiogenic shock. An intraaortic balloon pump had been inserted and she was breathing 100% O_2 through a nonrebreather mask. Blood gas set 1 in Table 10-9 was the 27th set of arterial blood gases from this patient.

Table 10-9 Arterial Blood-Gas Composition for Case 13

Set	pH	P_{CO_2} torr	$[HCO_3^-]$ mEq / L	P_{C_2} torr
1	7.36	33	18	112
2	7.30	39	19	39
3	7.38	31	18	149
4	7.44	32	21	112

In set 1 the patient had a compensated metabolic acidosis. At this point she had been breathing 100% O_2 from a nonrebreather mask for 18 hours. Her very high $F_{I_{O_2}}$ (1.0) and low arterial P_{O_2} indicate a large difference between alveolar and arterial P_{O_2}. Set 2 was obtained 40 minutes after set 1. At this point the very low arterial P_{O_2} indicated that adult respiratory distress syndrome (ARDS) had developed to the point that mechanical ventilation was imperative. When these results were obtained the patient was intubated and mechanically ventilated with 8 cm of positive end expiratory pressure (PEEP) and an $F_{I_{O_2}}$ of 0.6. Twenty minutes later set 3 was obtained. At this point the patient had a mixed metabolic acidosis and respiratory alkalosis, but her acid-base status and arterial oxygenation were in much safer ranges. Set 4 was obtained 2 hours after set 3 and 20 minutes after the $F_{I_{O_2}}$ was reduced to 0.4. The increasing alkalosis was indicative of improving oxygenation of her peripheral tissues, and the high arterial P_{O_2} indicated the effectiveness of the mechanical ventilation with PEEP. This patient survived the crisis and was discharged from the hospital three weeks later.

This is an excellent example of mismatching of blood flow and ventilation (Chapter 9). The extremely high $F_{I_{O_2}}$ in set 1 allowed even poorly ventilated regions to saturate hemoglobin with O_2. However, as lung damage progressed, many of the alveoli probably collapsed, creating a right to left shunt. Finally, even the high $F_{I_{O_2}}$ could not maintain arterial P_{O_2}, as shown in set 2. Note in set 2 that P_{CO_2} is still normal although it did increase over the previous value by 6 torr.

ARDS is a serious, often life-threatening response of lung tissue to a multitude of different kinds of insults. The patient in Case 13 developed the syndrome most likely because of cardiogenic shock, possible viral infection, and O_2 toxicity. It is the O_2 toxicity that could have been prevented if the very high $F_{I_{O_2}}$ had not been maintained so long. This patient improved rapidly after mechanical ventilation with lower $F_{I_{O_2}}$ was begun. The PEEP also contributed to better oxygenation of the blood, probably by inflating partially or totally collapsed alveoli, thereby reducing right to left shunting of blood. A complete discussion of the pathophysiology of ARDS and PEEP is beyond the scope of this text. The interested reader is directed to a text of respiratory pathophysiology for a more complete description of ARDS.

APPENDIX 1

p*Ka*

The dissociation constant of an acid, *Ka*, may be calculated if the concentration of H^+, conjugate base (B^-), and conjugate acid (*HB*) for a solution are measured or known. The relationship is shown in Equation A-1.1:

$$Ka = \frac{[H][B^-]}{[HB]} \qquad \text{(A-1.1)}$$

The stronger the weak acid, the greater the degree of dissociation of the acid and the larger the *Ka* will be. In the derivation of the Henderson-Hasselbalch equation, *Ka* was transformed to p*Ka*. The p*Ka* is related to *Ka* as shown in Equation A-1.2:

$$pKa = \log \frac{1}{Ka} \qquad \text{(A-1.2)}$$

Thus, the stronger the acid, the smaller the p*Ka* will be for that acid. The value of the p*Ka* can either be calculated from *Ka* or measured from the pH of a solution of a weak acid when the acid is 50% dissociated. The relationship is shown in Equations A-1.3 and A-1.4:

$$pH = pKa + \log \frac{[B^-]}{[HB]} \qquad \text{(A-1.3)}$$

At 50% dissociation, $[HB] = [B^-]$ and the ratio of $[B^-]/[HB] = 1$. The log of

1 = 0, and therefore, at 50% dissociation,

$$pH = pKa \tag{A-1.4}$$

When a weak acid is 50% dissociated and then titrated with a strong base, such as sodium hydroxide, or with a strong acid, such as HCl, a titration curve as shown in Figure A-1.1 can be obtained. Note that the curve is steepest when the acid is half titrated. At this midpoint, if either a strong acid or strong base is added, pH will change minimally. However, if strong acid or base were added at the extremes of these ratios, then larger pH changes would occur. In other words, maximum buffering occurs within 1 pH unit of the p*Ka* of a weak acid.

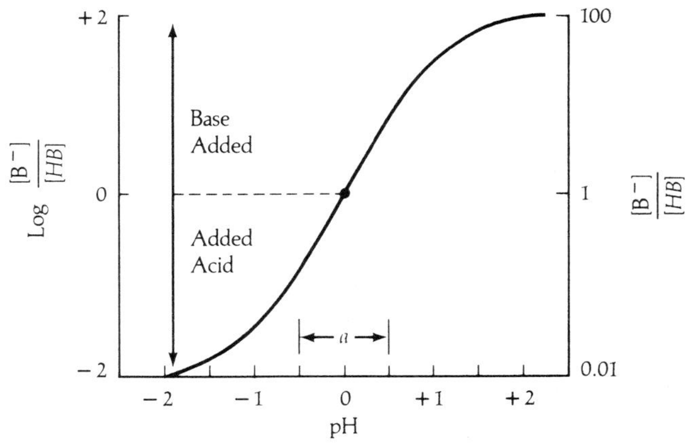

FIGURE A-1.1
Titration curve of a weak acid. The ordinate on the right is an exponential scale of the ratio of the concentration of conjugate base to conjugate acid. When the ratio is 1, the logarithm of the ratio is 0, as shown on the left ordinate. When the ratio is 1, pH = p*Ka*. The abscissa is scaled in units of difference in pH from the p*Ka*, that is, pH − p*Ka*. Maximum buffering (steepest part of curve) occurs at the p*Ka* and is nearly linear ±0.5 pH units from the value of p*Ka* as shown by the arrows (*a*). The curve appears to be nearly linear to ±1 pH unit from the value of p*Ka*, but when weak acids are titrated in the ranges of concentrations found in body fluids, maximum buffering occurs over a more narrow pH range.

Proteins such as hemoglobin, plasma proteins, and cell proteins have several weak acid groups or radicals as a part of their structure. Each of these groups tends to have a different p*Ka*. Hence, when proteins are titrated, each group buffers maximally within 1 pH unit of its p*Ka*. As a result, when acids are added to protein solutions in vivo or in vitro, maximum buffering is extended over a wider range than a single pH unit and the titration curve tends to be more linear. This is seen with the CO_2 buffer curve shown in Figure 5-11.

APPENDIX 2

Shunt Fraction

The derivation of this equation is based on the model shown in Figure A-2.1. In the model, cardiac output or total blood flow ($\dot{Q}_t$) is the sum of blood flow through pulmonary capillaries ($\dot{Q}_c$) in which gas exchange occurs plus the flow through the shunt ($\dot{Q}_s$) (Equation A-2.1). Blood flowing through the shunt and capillary paths mixes in the left ventricle and is ejected as mixed arterial blood.

$$\dot{Q}_t = \dot{Q}_c + \dot{Q}_s \qquad \text{(A-2.1)}$$

Therefore,

$$\dot{Q}_c = \dot{Q}_t - \dot{Q}_s \qquad \text{(A-2.2)}$$

The total amount of oxygen flowing to tissues per minute is the product of the total blood flow and oxygen concentration in arterial blood. That total must equal the sum of the total oxygen flowing from the capillaries plus that flowing from the shunt (Equation A-2.3).

$$\dot{Q}_t[O_2]_a = \dot{Q}_c[O_2]_c + \dot{Q}_s[O_2]_{mv} \qquad \text{(A-2.3)}$$

$[O_2]_c$ is oxygen concentration in blood flowing from pulmonary capillaries and $[O_2]_{mv}$ is the oxygen concentration in mixed venous blood flowing through the shunt. Rearranging the equation produces

$$\dot{Q}_s[O_2]_{mv} = \dot{Q}_t[O_2]_a - \dot{Q}_c[O_2]_c \qquad \text{(A-2.4)}$$

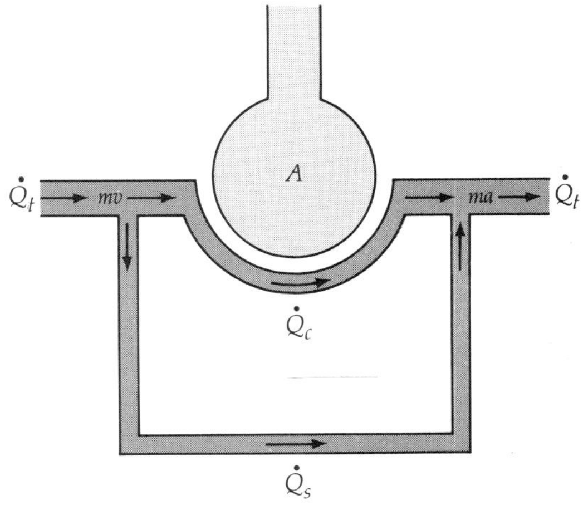

FIGURE A-2.1
Model used for calculation of shunt fraction. $\dot{Q}_t$ is total blood flow, $\dot{Q}_c$, is the blood flow through pulmonary capillaries and $\dot{Q}_s$ is the flow through the shunt. *A* represents alveoli. Mixed venous blocd (*mv*) flows into the pulmonary capillaries. The blood from the shunt and exchanging capillaries becomes mixed arterial blood (*ma*) after it is ejected from the left ventricle.

Substituting Equation A-2.2 into A-2.4

$$\dot{Q}_s[O_2]_{mv} = \dot{Q}_t[O_2]_a - (\dot{Q}_t - \dot{Q}_s)[O_2]_c \qquad \text{(A-2.5)}$$

$$\dot{Q}_s[O_2]_{mv} = \dot{Q}_t[O_2]_a - \dot{Q}_t[O_2]_c + \dot{Q}_s[O_2]_c \qquad \text{(A-2.6)}$$

Factoring and multiplying by (-1),

$$\dot{Q}_s([O_2]_c - [O_2]_{mv}) = \dot{Q}_t([O_2]_c - [O_2]_a) \qquad \text{(A-2.7)}$$

$$\frac{\dot{Q}_s}{\dot{Q}_t} = \frac{[O_2]_c - [O_2]_a}{[O_2]_c - [O_2]_{mv}} \qquad \text{(A-2.8)}$$

Equation A-2.8 is the fraction of the total blood flow that is shunted, or shunt fraction. This fraction may be calculated by obtaining both arterial and mixed venous blood samples and measuring the P_{O_2} of each. The concentration of O_2 in each may then be calculated from the percentage saturation and hemoglobin concentration (Chapter 3). The $[O_2]_c$ is usually estimated from the alveolar P_{O_2}, which is calculated from the alveolar gas equation (Appendix 3).

Frequently, to minimize errors, the patient is given 100% O_2 to breathe when shunt fraction is being calculated. At least 20 minutes of breathing this gas is required to ensure that the calculated and actual P_{O_2} of the alveolar space are the same. This time permits washout of much, but not all, of the nitrogen from the alveoli.

APPENDIX 3

Use of the Alveolar Gas Equation

It is sometimes necessary to calculate alveolar P_{O_2}. For example, the calculated alveolar P_{O_2} can be used to calculate the oxygen concentration in pulmonary capillary blood, which in turn is needed for calculating shunt fraction (Appendix 2, Equation A-2.8).

Equation 9-2 is really a shortened form of the alveolar gas equation:

$$P_{A_{O_2}} = P_{I_{O_2}} + \frac{P_{A_{CO_2}} \cdot F_{I_{O_2}}(1 - R)}{R} - \frac{P_{A_{CO_2}}}{R} \qquad \text{(A-3.1)}$$

In this equation, R is the respiratory exchange ratio that is shown in Equation A-3.2:

$$R = \frac{\dot{V}_{CO_2}}{\dot{V}_{O_2}} \qquad \text{(A-3.2)}$$

where $\dot{V}_{CO_2}$ is CO_2 excretion and $\dot{V}_{O_2}$ is oxygen uptake in liters per minute. In a steady state, CO_2 excretion equals CO_2 production, and this steady state is assumed in these equations. In most people who eat a balanced diet, the value of R is about 0.8. This means that CO_2 excretion is less than O_2 consumption and, therefore, the volume of gas excreted from the lungs per minute will be less than that inhaled per minute. The reason R is less than 1 is that some of the O_2 consumed is used to oxidize hydrogens on fats to water in addition to oxidizing the carbon to CO_2. Frequently, as a first approximation the middle term on the right-hand side of Equation A-3.1 is ignored. This leaves the alveolar gas equation in the form shown in Equation A-3.3, which is the same as Equation 9-2.

$$P_{A_{O_2}} = P_{I_{O_2}} - \frac{P_{A_{CO_2}}}{R} \qquad \text{(A-3.3)}$$

This approximation does not cause much error in the final calculated value of alveolar P_{O_2} (2–3 torr) if room air is the inspired gas. In fact, some authors simply add 2 torr to Equation 9-2 as a means of simplifying the equation. However, if very precise data are needed, then Equation A-3.1 must be used.

If $R = 1$, as can occur on a pure carbohydrate diet, Equation A-3.1 becomes

$$P_{A_{O_2}} = P_{I_{O_2}} - P_{A_{CO_2}} \qquad \text{(A-3.4)}$$

In Equation 4-9, it was shown that $P_{A_{CO_2}}$ is equal to a constant (K) divided by $\dot{V}_A$. When Equation 4-9 is substituted into Equation A-3.4,

$$P_{A_{O_2}} = P_{I_{O_2}} - \frac{K}{\dot{V}_A} \qquad \text{(A-3.5)}$$

This equation shows that if CO_2 production (metabolic rate) is constant, then as alveolar ventilation rate increases, alveolar P_{O_2} will increase, and vice versa. The increase in alveolar P_{O_2} is accomplished at the expense of alveolar P_{CO_2}; that is, at constant $\dot{V}_{CO_2}$, increased alveolar ventilation rate decreases alveolar P_{CO_2} and increases alveolar P_{O_2}.

Once alveolar P_{O_2} is calculated, the O_2 concentration in pulmonary capillary blood may be calculated as follows:

$$[O_2]_c = P_{A_{O_2}} \frac{(0.003 \text{ ml } O_2)}{\text{dl torr}} + \frac{\%\text{ Saturation}}{100}\left(\frac{1.39 \text{ ml } O_2}{\text{g Hb}}\right)[\text{Hb}] \qquad \text{(A-3.6)}$$

The percentage saturation may be obtained by assuming that $P_{A_{O_2}} = P_{O_2}$ in pulmonary capillary blood and then determining the percentage saturation from the O_2 saturation curve (see Figure 3-4) or from a nomogram. The nomogram is the more accurate method.

APPENDIX 4

Donnan Equilibrium

Donnan equilibrium is a special kind of equilibrium that exists across membranes separating nonpermeable ions in one solution from another solution that does not contain that ion. A model of such a situation is shown in Figure A-4.1. The membrane is not permeable to the protein anion (A^-). The model is analogous to capillaries separating plasma and ISF, in which side 1 is plasma and side 2 is ISF. Sodium concentration on side 1 is greater than that on side 2 (150 mEq/L plasma water versus 145 mEq/L in ISF). Chloride concentration on side 2 is greater than that on side 1.

At equilibrium, the *net* flux of Na^+ is zero and the *net* flux of chloride is also zero; there will be a potential difference across the membrane because of the unequal concentrations of cations and anions. This potential difference can be calculated from the Nernst equations, as shown in Equations A-4.1 and A-4.2:

$$E_{Na^+} = \frac{RT}{nF} \ln \frac{[Na^+]_1}{[Na^+]_2} \tag{A-4.1}$$

and

$$E_{Cl^-} = \frac{RT}{nF} \ln \frac{[Cl^-]_2}{[Cl^-]_1} \tag{A-4.2}$$

At equilibrium, the potential difference due to sodium (E_{Na^+}) and that due to

FIGURE A-4.1
Model for Donnan equilibrium. A large negatively charged anion, A^-, cannot cross the membrane, but both Na^+ and Cl^- can cross by diffusion. Side 1 could represent plasma in the capillary and side 2, ISF. In plasma, the protein is the negatively charged anion.

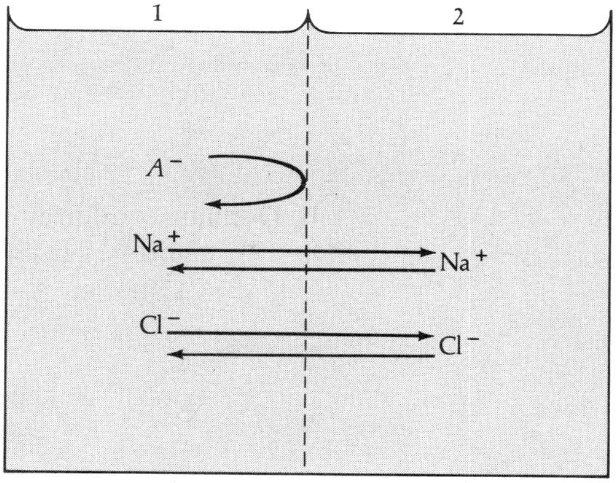

chloride (E_{Cl^-}) are equal. Therefore, since

$$E_{Na^+} = E_{Cl^-} \tag{A-4.3}$$

$$\frac{RT}{nF}\ln\frac{[Na^+]_1}{[Na^+]_2} = \frac{RT}{nF}\ln\frac{[Cl^-]_2}{[Cl^-]_1} \tag{A-4.4}$$

Therefore, at equilibrium,

$$\frac{[Na^+]_1}{[Na^+]_2} = \frac{[Cl^-]_2}{[Cl^-]_1} = r \tag{A-4.5}$$

where r is called the Donnan ratio.

It is important to note that Donnan equilibrium and osmotic equilibrium cannot exist simultaneously across the same membrane. This may be shown in the following general proof. If osmotic equilibrium were present in the model, *the ratio of the sums of the concentrations* of Na^+ and Cl^- on each side of the membrane would have to be less than 1, as shown in Equation A-4.6.

$$\frac{[Na^+]_1 + [Cl^-]_1}{[Na^+]_2 + [Cl^-]_2} < 1 \tag{A-4.6}$$

This is an implicit assumption, because if the ratio in Equation A-4.6 were equal to or greater than 1, total solute concentration on side 1 would be greater than that on side 2 because of the presence of the impermeant anion in plasma. In the model shown

$$[Na^+]_2 = [Cl^-]_2 \tag{A-4.7}$$

and from the Donnan relationship (Equation A-4.5)

$$[Na^+]_1 = r[Na^+]_2 \tag{A-4.8}$$

and

$$[\mathrm{Cl}^-]_1 = \frac{[\mathrm{Cl}^-]_2}{r} = \frac{[\mathrm{Na}^+]_2}{r} \qquad \text{(A-4.9)}$$

If Donnan equilibrium and osmotic equilibrium existed simultaneously, then $[Na^+]_2$ could be substituted for $[Cl^-]_2$, $r[Na^+]_2$ for $[Na^+]_1$, and $[Na^+]_2/r$ for $[Cl^-]_1$ in Equation A-4.6. The result is

$$\frac{r[\mathrm{Na}^+]_2 + [\mathrm{Na}^+]_2 / r}{[\mathrm{Na}^+]_2 + [\mathrm{Na}^+]_2} \qquad \text{(A-4.10)}$$

Therefore, Equation A-4.10 becomes

$$\frac{r + 1/r}{2} = \frac{r^2 + 1}{2r} \qquad \text{(A-4.11)}$$

Completing the square by adding and subtracting $2r$ leaves

$$\frac{(r^2 - 2r + 1) + 2r}{2r} = \frac{(r-1)^2}{2r} + 1 \qquad \text{(A-4.12)}$$

There are three possibilities:

1. If the large ion had no net charge, $r = 1$, and Equation A-4.12 $= 1$.
2. If the large ion has a positive charge, $r > 1$, and Equation A-4.12 > 1.
3. If the large ion has a negative charge, $r < 1$, but Equation A-4.12 is *still* > 1.

All three possibilities violate the requirement of osmotic equilibrium shown in Equation A-4.6. Therefore, osmotic equilibrium cannot exist across the same membrane simultaneously with Donnan equilibrium.

The ISF is in Donnan equilibrium with plasma across the capillary membrane when filtration rate equals osmotic reabsorption (midpoint where the hydrostatic and osmotic pressure lines intersect in Figure 2-4). At that point, the hydrostatic pressure in the capillary opposes osmotic pressure to prevent osmosis and the generation of osmotic equilibrium.

Red blood cells are in osmotic equilibrium with plasma water and they have a large impermeant anion in the ICF. Prevention of Donnan equilibrium is achieved by means of an active transport pump, which extrudes sodium from the cell interior. Cells in general must prevent Donnan equilibrium or Na^+ will enter the cell accompanied by osmosis of water and the cells will lyse. One of the effects of hypoxia is diminishing ATP necessary for active transport of Na^+ out of cells and the cells begin to swell. If hypoxia is severe, swelling can lead to lysis and death.

Estimating Osmolality of Plasma

APPENDIX

5

Osmolality and osmolarity are terms designating total solute concentration of a solution. These terms are especially useful with solutions containing several different kinds of solutes. An osmole is 1 Avogadro number (6.02×10^{23}) of molecules or solute particles. A 1-osmolal solution contains 1 Avogadro number of solute particles (or 1 mole of solute particles) *per kilogram of solvent*. A 1-osmolar solution contains 1 Avogadro number of solute molecules or particles *per liter of solution*. Body fluids contain less than 1 Osm/L and, therefore, total solute concentration is usually given in milliosmoles rather than osmoles.

The osmolality of plasma is, on the average, about 290 mOsm/kg plasma water and the normal range varies from 285 to 295 mOsm/kg plasma water. Both crystalloid and colloid solutes contribute to the osmolality. The most ubiquitous solute is sodium, which contributes approximately 48% or 140 mOsm to the total. The anions accompanying sodium (Cl^- and HCO_3^-) add another 125 to 130 mOsm and unmeasured anions also contribute a few more milliosmoles.

Osmolality of sodium or plasma can be measured in the laboratory, and it can also be estimated from the concentrations of electrolytes and nonelectrolytes measured in serum. Equation A-5.1 illustrates the estimation of osmolality from serum concentrations of Na^+, glucose, and BUN:

$$\text{Estimated osmolality} = [Na^+] \times 2 + \frac{[\text{Glucose}]}{18} + \frac{[\text{BUN}]}{2.8} \qquad \text{(A-5.1)}$$

The units in the different parts of Equation A-5.1 are not dimensionally the same, so the equation is not proper in the mathematical sense. The sodium concentration is multiplied by 2 so that anions accompanying sodium are taken into account. The serum glucose concentration is divided by 18. The number obtained is equal to the millimoles of glucose per liter of serum. To be dimensionally correct, the term, [Glucose]/18, should be

$$\frac{\text{mg Glucose / dl}}{\text{180 mg / mMol}} \times \frac{\text{10 dl}}{\text{L}} = \frac{\text{mMol Glucose}}{\text{L}} \tag{A-5.2}$$

The BUN is really the urea nitrogen concentration in serum. Its units are milligrams of urea nitrogen per deciliter of serum. Nitrogen is 28/60 of the urea molecule, that is, from the chemical formula for urea, N_2H_4CO, the total molecular weight is 60 and nitrogen constitutes almost half. To be dimensionally correct, the last term in Equation A-5.1 should be

$$\frac{\text{mg urea nitrogen / dl}}{\text{28 mg / mMol}} \times \frac{\text{10 dl}}{\text{L}} = \frac{\text{mMol urea nitrogen}}{\text{L}} \tag{A-5.3}$$

Actually 1 mMol of urea nitrogen is equivalent to 1 mMol of urea. There is no really good reason today for using the archaic method of reporting urea concentration as urea nitrogen, but it is tradition and it will probably remain.

The number that is calculated from Equation A-5.1 is really an estimate of osmolarity of serum because all of the terms are based on concentration of solute per liter, not per kilogram of serum water. This is in contrast to the results from laboratory analysis, which provides true osmolality. The number obtained from Equation A-5.1 is arbitrarily given the units of osmolarity.

In a healthy person, the $[Na^+]$ of serum would be approximately 140 mEq/L, the [Glucose] would be about 90 mg/dl, and the [BUN] about 15 mg/dl. From Equation A-5.1 the estimated osmolarity would be

$$(140 \times 2) + (90 / 18) + (15 / 2.8) = 290 \text{ mOsm / L} \tag{A-5.4}$$

Sometimes, as a shortcut, serum osmolarity is estimated by simply doubling the sodium concentration, that is, using only the first term in Equation A-5.1. In the preceding example, such a shortcut yields an estimated osmolarity of 280 mOsm/L, which is about a 3% difference from the more cumbersome approach of Equation A-5.1. Many clinicians simply ignore the small error. However, there can be danger in such a shortcut. For example, if the glucose and BUN concentrations are ignored in the first set of serum concentrations shown in Table 10-6 for Case 8, serum osmolality would be grossly underestimated. Twice the $[Na^+]$ provides an estimate of 268 mOsm/L, whereas Equation A-5.1 yields 330 mOsm/L. Because the patient was markedly dehydrated, the first estimate could inadvertently lead to the conclusion that she had a hypoosmotic volume contraction when in fact she had hyperosmotic volume contraction. Thus, it is not necessarily a good idea to estimate serum osmolality from serum $[Na^+]$ alone unless it is known for certain that other solute concentrations are normal. Care should also be taken to avoid estimating serum osmolality using either Equation A-5.1 or the shortcut method when serum lipid concentration is markedly elevated (pseudohyponatremia).

Hyponatremia in the presence of hyperosmolality can occur with redistribution of water. The most common example of this cause of hyponatremia is in hyperglycemia from diabetes mellitus. The marked increase in glucose concentration in the ECF causes osmosis of water from the ICF, which dilutes the Na^+. However, total osmolality, as shown previously, may be increased. The decrement in $[Na^+]$ caused by osmotic shift of water can be calculated. Generally, $[Na^+]$ decreases by 1.5 mEq/L for every 100 mg/dl increase in glucose concentration because of water shift (see Figure 8-1*A*).

Let's take a hypothetical situation. Suppose a 50-kg male developed diabetes and his glucose concentration in his serum was 1000 mg/dl. As a first approximation, assume that the glucose is added to only the ECF and that there has been no net water loss from the body. Also assume that, before the glucose concentration increased above normal, serum osmolarity was 300 mOsm/L.

In a 50-kg male, 20% of body weight is ECF and 40% is ICF. In this individual, initial ECF volume would be 10 L and ICF volume would be 20 L. A glucose concentration of 100 mg/dl is 5.55 mOsm/L. Therefore, a glucose concentration 900 mg/dl greater than normal would add 49.95 mOsm/L to his ECF. Total osmolarity of ECF should then be (300 mOsm/L + 49.95 mOsm/L) × 10 L = 3499.5 mOsm in his ECF. Because no change is assumed to have occurred in the solute content of ICF, he still has 6000 mOsm in the ICF (300 mOsm/L × 20 L). Because the glucose increases the osmolality of the ECF, water will shift from ICF to ECF. The amount shifted will increase the volume of the ECF by *X* L and decrease the volume of the ICF by the same amount. Thus

$$\frac{3499.5\ \text{mOsm}}{10\ \text{L} + X\ \text{L}} = \frac{6000\ \text{mOsm}}{20\ \text{L} - X\ \text{L}} \qquad \text{(A-5.5)}$$

$$69{,}990 - 3499.5X = 60{,}000 + 6000X \qquad \text{(A-5.6)}$$

$$9499.5X = 9990 \qquad \text{(A-5.7)}$$

$$X = 1.05 \qquad \text{(A-5.8)}$$

Therefore, the new osmolarity of ECF is

$$\frac{3499.5\ \text{mOsm}}{11.05\ \text{L}} = 316.7\ \text{mOsm/L} \qquad \text{(A-5.9)}$$

which is also the new osmolarity of the ICF.

Prior to the hyperglycemia, the $[Na^+]$ was normal at 140 mEq/L. That means that in his normal 10 L of ECF he had 1400 mEq of sodium. Now that sodium is dissolved in 11.05 L, which means that sodium concentration has decreased:

$$\frac{1400 \text{ mEq Na}^+}{11.05 \text{ L}} = 126.7 \text{ mEq/L} \qquad \text{(A-5.10)}$$

The $[Na^+]$ has decreased from a normal of 140 mEq/L to 126.7 mEq/L or a decrease of 13.3 mEq/L. Therefore, for a 900 mg/dl increase in glucose concentration, sodium concentration decreases by 13.3 mEq/L. Thus, for every 100 mg/dl increase in glucose concentration, one may expect that sodium concentration would decrease by about 1.5 mEq/L.

One must also keep in mind the assumptions used in making this estimate. It was assumed that no sodium or water was lost, therefore, the change in $[Na^+]$ calculated represents only dilution of sodium. In fact, sodium is usually lost as a part of the osmotic diuresis produced by glucosuria. The $[Na^+]$ in a patient's ECF may not be as low as that predicted, and this is shown in Case 8 in Chapter 10. From the glucose concentration 958 mg/dl, the sodium concentration in serum would have been 125.6 mEq/L ($140 - 1.5 \times 9.58$) if water had not been excreted by the kidneys. The $[Na^+]$ in the patient's serum was actually 134 mEq/L.

The usefulness of the calculation becomes questionable if hyperglycemia has been present for a long time. However, if hyperglycemia is acute, then this simple calculation permits a quick estimate of the expected decrease in $[Na^+]$ based on redistribution of water. If $[Na^+]$ is less than predicted, then there is more dilution than can be accounted for by redistribution of water. If the $[Na^+]$ is greater than predicted, then water has been lost from the ECF compartment, usually through the kidneys or insensible loss.

The Anion Gap

APPENDIX

6

It is well known from the principle of electroneutrality that the number of milliequivalents of cations in serum is balanced by an equal number of milliequivalents of anions. Therefore, the sum of all cations in a liter of serum (measured in milliequivalents) must equal the sum of all anions measured in milliequivalents:

$$\Sigma\text{mEq cations} / \text{L} = \Sigma\text{mEq anions} / \text{L} \qquad \text{(A-6.1)}$$

The number of milliequivalents of cations per liter is $[Na^+] + [K^+] + [Mg^{++}] + [Ca^{++}] + [\text{other cations}]$. The number of milliequivalents of anions per liter is $[Cl^-] + [HCO_3^-] + [HPO_4^=] + [SO_4^=] + [\text{protein}^-] + [\text{other conjugate base anions}]$. Ordinarily, only the concentrations of Na^+, K^+, Cl^-, and HCO_3^- are measured routinely.* The rest are considered unmeasured cations (UC) and unmeasured anions (UA). In these terms Equation A-6.1 becomes

$$[Na^+] + [K^+] + [UC] = [Cl^-] + [HCO_3^-] + [UA] \qquad \text{(A-6.2)}$$

Rearranging

$$[Na^+] + [K^+] - [Cl^-] - [HCO_3^-] = [UA] - [UC] \qquad \text{(A-6.3)}$$

*The $[HCO_3^-]$ is actually calculated, not measured.

Normally, the concentration of unmeasured cations is about 7 mEq/L and the concentration of unmeasured anions is about 23 mEq/L. The difference in concentration between unmeasured anions and unmeasured cations is called the anion gap, which is equal to the difference in concentrations of measured cations and measured anions.

If $[Na^+]$ is 140 mEq/L, $[Cl^-]$ is 103 mEq/L, $[K^+]$ is 4 mEq/L, and $[HCO_3^-]$ is 25 mEq/L, then the anion gap is 16 mEq/L and is normal. The value of the anion gap is sometimes calculated without $[K^+]$; thus, the value of $[K^+]$ is lumped with [UC]. If this is done using the concentrations given earlier, the anion gap would be 12 mEq/L. Generally, the anion gap can vary 4 mEq/L from those average values and still be considered normal. Care must be taken when interpreting a reported anion gap. If the reported value is 18 mEq/L, it is normal if Equation A-6.3 is used but increased above normal if $[K^+]$ is left out of the calculations.

INCREASED ANION GAP

The anion gap increases most commonly because of an increase in the concentration of unmeasured anions. For example, when lactic acid is added to the blood during tissue hypoxia, the hydrogen ions dissociated from the lactic acid are buffered using up HCO_3^- as shown in Equation 6-1 of Chapter 6. The conjugate base of the added acid (lactate) in effect replaces the HCO_3^-, and therefore, increases unmeasured anion concentration. A similar scenario can be shown for ketoacids produced from lipid metabolism in diabetic ketoacidosis. The anion gap always increases during metabolic acidosis if the cause of the acidosis is from addition of fixed acids that have nonchloride conjugate bases. These conjugate bases include but are not limited to formate from methanol intoxication, salicylate from aspirin poisoning, lactate, acetoacetate from keto-acids, and unidentified anions produced from ethylene glycol poisoning. Finally, the anion gap will also be increased if there is a laboratory error in which the measured cation concentration (Na^+ or K^+) is reported at falsely high values. Pseudohyperkalemia can also cause a small increase in the anion gap.

NORMAL ANION GAP

The anion gap will remain within the normal range if an acidosis is caused by addition of chloride-containing acid such as HCl or substances that are converted to HCl by metabolism such as NH_4Cl, arginine chloride, or lysine chloride. It will also be normal in the metabolic acidosis caused by decreased renal generation of HCO_3^- such as occurs in renal tubular acidosis and renal disease. In early stages of renal failure, anion gap may be normal until excretion of unmeasured anions becomes impaired.

Anion gap will usually be normal with metabolic acidosis from any process that causes loss of HCO_3^- and retention of chloride such as intestinal loss of HCO_3^- with diarrhea or ureterosigmoidostomy. Chloride is absorbed in exchange for the secretion of HCO_3^- by the mucosa of the distal small bowel and colon. The large volume of luminal fluid associated with diarrhea provides a sink for HCO_3^- secretion. Transplanting the ureters to the sigmoid colon causes absorption of Cl^- from the urine and secretion of HCO_3^-.

If acidosis is caused by H^+–K^+ exchange or H^+–Na^+ exchange across cell membranes, the H^+ is buffered with HCO_3^- and, therefore, measured anion concentration decreases. Since Na^+ and K^+ are taken up by the cells in this process of exchange, measured cation concentration also decreases; hence, there will generally be no change in the anion gap. In these types of metabolic acidosis in which anion gap is normal, the chloride concentration is increased. When chloride concentration is greater than normal, these kinds of acidotic disturbances are referred to as hyperchloremic metabolic acidosis.

DECREASED ANION GAP

Decreased anion gap is less common than increased anion gap. From equation A-6.3 it is clear that an increase in unmeasured cation concentration such as occurs in hypercalcemia or hypermagnesemia will diminish the magnitude of the anion gap. It should also be noted that if $[K^+]$ is not included in the calculation, that is, if K^+ is considered as part of the unmeasured cations, then hyperkalemia can also cause a reduced anion gap.

USE OF ANION GAP

The anion gap is useful in obtaining data to support an assessment or diagnosis. For example, if increased production of lactic acid is suspected as the cause of a metabolic acidosis, then the anion gap should be increased because of an increase in the unmeasured anion concentration. If anion gap is normal or decreased, this particular assessment should be questioned. Calculation of the anion gap provides an adjunct to assessment that helps to confirm or refute suspected causes of electrolyte and acid-base disturbances.

APPENDIX 7

Base Excess and Deficit

BUFFER BASES IN BODY FLUIDS

Base excess and deficit are terms used to identify the presence of an excess or deficit of *buffer bases* compared to normal. Buffer bases* are the conjugate bases of weak acids found in body fluids. These bases plus their conjugate acids constitute the buffer systems found in the body. The major buffer bases in body fluids are shown in Table A-7.1.

The total amount of *nonbicarbonate* buffer bases (items 2–5) present in body fluids at any given moment depends on the total amount of each buffer present and the pH of the respective compartments. In general, the nonbicarbonate buffer bases are not subject to rapid regulation for acid-base purposes, and hence, the total amount of each nonbicarbonate buffer system (conjugate acid plus conjugate base) is fixed at any one time.

The $CO_2 - H_2CO_3 - HCO_3^-$ buffer system is unique in that both the P_{CO_2} (conjugate "acid") and $[HCO_3^-]$ (conjugate base) are regulated by the respiratory system and kidneys, respectively. Because the $[HCO_3^-]$ in the ECF is regulated, the total buffer base content of the body, β^- (in equation A-7.1), can be altered by changing the HCO_3^- content of body fluids.

$$\beta^- = (HCO_3^- + Pr_p^- + HPO_4^= + Pr_c^- + Hb^-) \qquad \text{(A-7.1)}$$

*In the physiologic pH range, all buffer bases are anions.

Table A-7.1 Buffer Bases in Body Fluids

1. HCO_3^-
2. Plasma proteinate (Pr_p^-)
3. Cell proteinate (Pr_c^-)
4. Cell phosphate ($HPO_4^=$)
5. Hemoglobin (Hb^-)

where Pr_p^- is total plasma proteinate, $HPO_4^=$ is *cell* phosphate, Pr_c^- is cell proteinate, and Hb^- is hemoglobinate in blood. The concentrations of other buffer bases, notably $HPO_4^=$ in the ECF, are too low to be of major importance and are, therefore, ignored in this discussion. However, it should be remembered that $HPO_4^=$ is an important buffer base in cells.

Each buffer base is in equilibrium with its respective conjugate acid. The total buffer base equilibrium may be simplified as shown in Equation A-7.2:

$$H\beta \rightleftharpoons H + \beta^- \qquad \text{(A-7.2)}$$

The total amount of buffer base in the body at normal pH and temperature is the reference for all measurements of base excess and deficit. The normal value of base excess is arbitrarily set at zero; in other words, in health at normal pH and body temperature, there is neither an excess nor a deficit of buffer base. In acid-base disturbances, base excess (BE) may be thought of as the difference between what is measured and the normal value (Equation A-7.3).

$$BE = \begin{pmatrix} \text{Measured} \\ \text{buffer base} \\ \text{concentration} \end{pmatrix} - \begin{pmatrix} \text{Normal} \\ \text{buffer base} \\ \text{concentration} \end{pmatrix} \qquad \text{(A-7.3)}$$

For example, suppose $[H^+]$ of body fluids increases from addition of fixed acids. The increased association of H^+ and β^- resulting from this addition (Equation A-7.2) decreases total buffer base content by converting β^- to its conjugate acid ($H\beta$). As a result, the difference between measured buffer base content and normal will be negative (a negative base excess) and a base deficit will be present. On the other hand, if a strong base such as sodium hydroxide (NaOH) is added to body fluids, the $[H^+]$ decreases and the dissociation of the conjugate acids buffers the added base. The net effect is to increase the concentration of β^-. The difference between measured buffer base concentration and normal would then be positive, and a base excess would be present. Therefore, when total body buffer base increases above normal, a base excess is present, and when total body buffer base decreases, a base deficit is present. A note of caution—just because the $[HCO_3^-]$ of ECF changes does not mean that $[\beta^-]$ changes. This point is discussed in more detail below.

DISTURBANCES CAUSING BASE EXCESS

Addition of base such as the salt of a weak acid or by generation of excess HCO_3^- by the kidney or gastric mucosa will add buffer base to the body fluids. The reaction is summarized in Equation A-7.4.

$$\begin{array}{ccc} & B^- & \\ & + & \\ H\beta \rightleftharpoons & H & + \beta^- \\ & \upharpoonleft\downharpoonright & \\ & HB & \end{array} \tag{A-7.4}$$

B^- is the added base. Addition of the base causes dissociation of the conjugate acids of the buffer bases ($H\beta$) and increases total buffer base content (β^-) as well as the concentration of these buffer bases in body fluids. All of the buffering reactions in each of the fluid compartments of the body are summarized in Figure 5-9*B*. Therefore, any chemical reaction causing the dissociation of the conjugate acids of buffer bases will tend to generate a base excess. These reactions occur during generation of a metabolic alkalosis and with compensation of a respiratory acidosis. *In other words, any time the blood-gas composition falls within areas I, II, III, or IV of Figure 6-4, a base excess is present.*

DISTURBANCES CAUSING BASE DEFICIT

Addition of a fixed acid, such as ketoacids and lactic acid, or loss of HCO_3^- through reduced generation by the kidneys or loss through the GI system will decrease total buffer base in the body. The reaction is summarized in Equation A-7.4, in which addition of fixed acid (HB) adds H^+ to body fluids. The H^+ is buffered by the buffer bases (β^-) by association to form their conjugate acids ($H\beta$). The net effect is decreased concentration of buffer bases in body fluids, which is summarized in Figure 5-9*A*. Thus, any chemical reaction that causes association of buffer bases with H^+ will tend to produce a base deficit. These reactions occur during the generation of metabolic acidosis and during compensation for a respiratory alkalosis. *In other words, any time blood-gas composition falls into regions V, VI, VII, or VIII of Figure* 6-4, *a base deficit is present.*

DISTURBANCES CAUSING NEITHER BASE EXCESS NOR DEFICIT

When CO_2 is added to body fluids, a series of buffering reactions, shown in Figure 5-4*A*, takes place in the body fluid compartments. The H^+ dissociating from H_2CO_3 is buffered by the nonbicarbonate buffer bases only (Table A-7.1, items 2–5). These buffering reactions cause HCO_3^- to be generated as summarized in Equation A-7.5:

$$\begin{array}{c} CO_2 + H_2O \rightleftharpoons H_2CO_3 \\ \qquad\qquad\quad \upharpoonleft\downharpoonright \\ \qquad\qquad\quad H\beta' \rightleftharpoons H^+ + \beta'^- \\ \qquad\qquad\quad + \\ \qquad\qquad\quad HCO_3^- \end{array} \qquad (A\text{-}7.5)$$

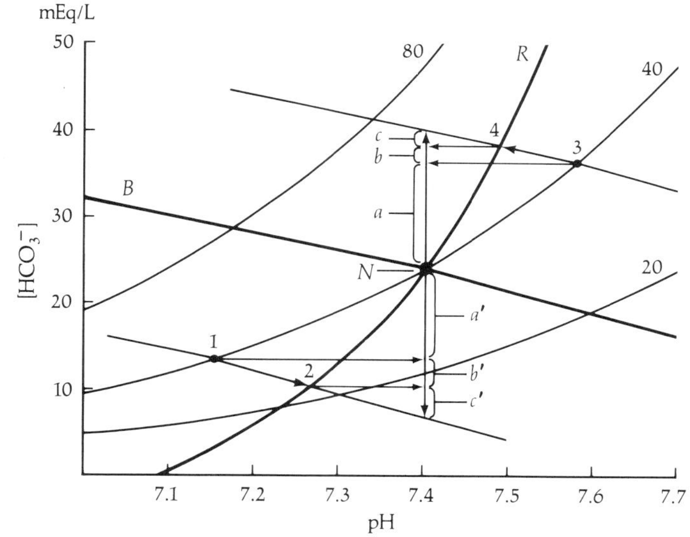

FIGURE A-7.1 Estimation of base excess and deficit. *R* is the respiratory compensation curve. Base deficit is present when blood-gas composition is below the CO_2 buffer curve (*B*). For example, addition of fixed acids without compensation (point 1) incurs a base deficit. The contribution of extracellular HCO_3^- and plasma proteins to the buffering is shown by *a′*. The actual base deficit is greater than the change in [HCO_3^-] shown by *a′* because of the presence of other buffers. The change in [HCO_3^-] caused by compensation (shift from point 1 to 2) is shown by distance *b′*. No change in base deficit occurs with compensation because the compensation is caused by loss of CO_2 alone. (A small contribution due to lactic acid production with hypocapnea may be present but will amount to only 1–3 mEq / L.) For every milliequivalent of HCO_3^- lost in compensation, another buffer base is generated, as described in the text. The estimated magnitude of the base deficit is the sum of $a' + b' + c'$, which is the vertical distance from the normal point, *N*, to the new position of the CO_2 buffer curve passing through points 1 and 2. The sum of *b′* and *c′* represents the contribution of nonbicarbonate buffers.

Base excess is present when blood-gas composition is above the CO_2 buffer curve (*B*). For example, if a base is added to body fluids, an uncompensated metabolic alkalosis will result, as shown by point 3. Respiratory compensation shifts the composition from point 3 to point 4. Distance *a* represents the contribution of carbonic acid and plasma proteins to the buffering. Distance *b* is the increase in [HCO_3^-] caused by compensatory CO_2 retention and will incur *no change* in the base excess. The base excess is estimated by the vertical distance $a + b + c$, which represents the parallel shift of the CO_2 buffer curve in response to the base excess.

It should be noted that actual base excess and deficit will be somewhat different from this estimate because the shifts in the CO_2 buffer curve with base excess and deficit are not quite parallel with the normal curve. The estimate is accurate enough for clinical purposes.

β'^- is the symbol for nonbicarbonate buffer bases. In this particular reaction, every *non*bicarbonate buffer base helps to buffer the H^+ from H_2CO_3. However, for every nonbicarbonate buffer base removed by association with H^+ to form its conjugate acid, a new HCO_3^- is formed. Thus, total buffer base, β^-, does not change, nor does the total concentration of buffer base change. A similar argument could be made for loss of CO_2 (Figure 5-4*B*) if the only events were those shown in Figure 5-4. Those changes are shown as the dashed extension of the CO_2 buffer curve in Figure 5-11. However, hypocapnea causes generation of lactic acid and, consequently, additional loss of HCO_3^-. Therefore, when volatile acid concentration increases, and as long as compensation does not occur, there will be no change in buffer base concentration. With loss of CO_2, if only buffering occurs, then no base deficit will occur (dashed line, Figure 5-11 and Figure 5-4*B*). Addition of lactic acid from hypocapnea causes a rather mild base deficit to be generated. This has to be remembered when interpreting blood-gas data obtained from patients who are in a primary respiratory alkalosis.

There is a change in buffer base content with compensation for respiratory acid-base disturbances. When compensation occurs, base excess will be present in compensated respiratory acidosis, and a base deficit will be present in compensated respiratory alkalosis.

The concepts of base excess and deficit were developed to help quantify the metabolic component in acid-base disturbances. Usually, the magnitude of the base excess or deficit will be reported along with the results of blood-gas analysis from the laboratory. Base excess and deficit can help to identify the severity of the disorder and sometimes may help in determining whether or not a patient has a compensated acid-base disorder or a complex type of disturbance. Some physicians use the magnitude of the base deficit to calculate the amount of HCO_3^- needed for treating metabolic acidosis. However, caution must be exercised, because replacement of an entire deficit is fraught with danger for the patient because of slow equilibration of HCO_3^- across the blood-brain barrier. (See Chapter 4 and also the analysis of Case 8 in Chapter 10.) A simple way to estimate base excess or deficit is shown in Figure A-7.1.

Additional Readings/References

CHAPTER 1
Fundamental Concepts Central to Fluid, Electrolyte, and Acid-Base Regulation

1. Berne, R. M. and M. N. Levy. *Cardiovascular Physiology*, 4th ed. C. V. Mosby Co., St. Louis, 1981. Chapter 3 gives an excellent discussion of bulk flow.
2. Carveth, M. L., A. J. Schriver, W. E. Peterson, and J. L. Keyes. "Acid-Base Assessment with Peripheral Venous Blood." *Heart and Lung* 13:48–54, 1984.
3. Cohen, J. J. and J. P. Kassirer. *Acid-Base*. Little, Brown, and Co., Boston, 1982. This is an excellent source for more advanced reading on clinical acid-base regulation.
4. Davenport, H. W. *The ABC of Acid-Base Chemistry*, 6th ed. University of Chicago Press, Chicago, 1974. An excellent little monograph that is considered a classic on the subject. Davenport popularized the pH/[HCO_3^-] diagram.
5. Flickinger, C. J., J. C. Brown, H. C. Kutchai, and J. W. Ogilvie. *Medical Cell Biology*. W. B. Saunders Co., Philadelphia, 1979.
6. Henderson, L. J. *Blood: A Study in General Physiology*. Yale University Press, New Haven, 1928. A classic monograph well worth study. Many of the concepts used today are from or have been modified from this treatise. This is an excellent work for graduate students who want a more complete historical and quantitative treatment of the subject.
7. Hills, A. G. *Acid-Base Balance*. Williams & Wilkins Co., Baltimore, 1973. A scholarly detailed work with numerous references, but this monograph is not for the beginning student.
8. Keyes, J. L. "Basic Mechanisms Involved in Acid-Base Homeostasis." *Heart and Lung* 5:239–246, 1976.

9. Lassiter, W. E. and C. W. Gottschalk. "Volume and Composition of Body Fluids," in V. B. Mountcastle (ed.), *Medical Physiology*, 14th ed., Volume II. C. V. Mosby Co., St. Louis, 1980.
10. Murray, J. F. *The Normal Lung*. W. B. Saunders Co., Philadelphia, 1976.
11. Pitts, R. F. *Physiology of the Kidney and Body Fluids*, 3rd ed. Yearbook Medical Publishers, Chicago, 1974.
12. Sullivan, L. P. and J. J. Grantham. *Physiology of the Kidney*, 2nd ed. Lea & Febiger, Philadelphia, 1982.
13. Vander, A. J., J. H. Sherman, and D. S. Luciano. *Human Physiology*, 3rd ed. McGraw-Hill Book Co., New York, 1980. This is an outstanding introductory text to human physiology. The discussion is extremely clearly presented.
14. Woodbury, D. M. "Physiology of Body Fluids," in T. C. Ruch and H. D. Patton (eds.), *Physiology and Biophysics*, 20th ed., Volume II. W. B. Saunders Co., Philadelphia, 1974.

CHAPTER 2
Fluid and Osmolality Balance: Normal Homeostatic Processes

1. Berne, R. M. and M. N. Levy. *Cardiovascular Physiology*, 4th ed. C. V. Mosby Co., St. Louis, 1981. Chapter 6 is an excellent review of capillary dynamics.
2. Lassiter, W. E. and C. W. Gottschalk. "Volume and Composition of Body Fluids," in V. B. Mountcastle (ed.), *Medical Physiology*, 14th ed., Volume II. C. V. Mosby Co., St. Louis, 1980.
3. Pitts, R. F. *Physiology of the Kidney and Body Fluids*, 3rd ed. Yearbook Medical Publishers, Chicago, 1974. Chapter 2 is especially pertinent.
4. Rose, B. D. *Clinical Physiology of Acid-Base and Electrolyte Disorders*. McGraw-Hill Book Co., New York, 1977.
5. Valtin, H. *Renal Function*, 2nd ed. Little, Brown, and Co., Boston, 1983.
6. Woodbury, D. M. "Physiology of Body Fluids," in T. C. Ruch and H. D. Patton (eds.), *Physiology and Biophysics*, 20th ed., Volume II. W. B. Saunders Co., Philadelphia, 1976.

CHAPTER 3
Blood Gases and Blood-Gas Transport

1. Comroe, J. H., Jr. *Physiology of Respiration*, 2nd ed. Yearbook Medical Publishers, Chicago, 1974.
2. Davenport, H. W. *The ABC of Acid-Base Chemistry*, 6th ed. University of Chicago Press, Chicago, 1974.
3. Keyes, J. L. "Blood-Gases and Blood-Gas Transport." *Heart and Lung* 3:945–954, 1974.
4. Levitsky, M. G. *Pulmonary Physiology*. McGraw-Hill Book Co. New York, 1982.
5. Murray, J. F. *The Normal Lung*. W. B. Saunders Co., Philadelphia, 1976.
6. Roughton, F. J. W. "Transport of Oxygen and Carbon Dioxide," in *Handbook of Physiology*, *Respiration*, Volume I, pp. 767–828. American Physiological Society, Washington, D.C., Williams & Wilkins Co., Baltimore, 1964.
7. Slonim, N. B. and L. H. Hamilton. *Respiratory Physiology*, 4th ed. C. V. Mosby Co., St. Louis, 1981.

8. West, J. B. *Respiratory Physiology, The Essentials*, 2nd ed. Williams & Wilkins Co., Baltimore, 1979.

CHAPTER 4
Regulation of P_{CO_2} and Bicarbonate Concentration of Plasma

1. Berger, A. J., R. A. Mitchell, and J. W. Severinghaus. "Regulation of Respiration," *New England Journal of Medicine* 297:92–97, 138–143, 194–201, 1977.
2. Comroe, J. H., Jr. *Physiology of Respiration*, 2nd ed. Yearbook Medical Publishers, Chicago, 1974.
3. Davis, J. O. and R. H. Freeman. "Mechanisms Regulating Renin Release." *Physiological Review* 56:1, 1976.
4. Keyes, J. L. "Basic Mechanisms Involved in Acid-Base Homeostasis," *Heart and Lung* 5: 239–246, 1976.
5. Kryger, M. H. *Pathophysiology of Respiration*. John Wiley & Sons, New York, 1981.
6. Levitsky, M. G. *Pulmonary Physiology*. McGraw-Hill Book Co., New York, 1982.
7. Murray, J. F. *The Normal Lung*. W. B. Saunders Co., Philadelphia, 1976.
8. Pitts, R. F. *Physiology of the Kidney and Body Fluids*, 3rd ed. Yearbook Medical Publishers, Chicago, 1974.
9. Slonim, N. B. and L. H. Hamilton. *Respiratory Physiology*, 4th ed. C. V. Mosby Co., St. Louis, 1981.
10. Sullivan, L. P. and J. J. Grantham. *Physiology of the Kidney*, 2nd ed. Lea & Febiger, Philadelphia, 1982.
11. Valtin, H. *Renal Function*, 2nd ed. Little, Brown, and Co., Boston, 1983.
12. Vander, A. J. *Renal Physiology*, 2nd ed. McGraw-Hill Book Co., New York, 1980.

CHAPTER 5
Buffering of Volatile and Nonvolatile Acids and Bases

1. Cohen, J. J. and J. P. Kassirer. *Acid-Base*. Little, Brown, and Co., Boston, 1982.
2. Davenport, H. W. *The ABC of Acid-Base Chemistry*, 6th ed. University of Chicago Press, Chicago, 1974.
3. Giebisch, G., L. Berger, and R. F. Pitts. "The Extrarenal Response to Acute Acid-Base Disturbances of Respiratory Origin." *Journal of Clinical Investigation* 34:231–245, 1955.
4. Henderson, L. J. *Blood: A Study in General Physiology*. Yale University Press, New Haven, 1928.
5. Hills, A. G. *Acid-Base Balance*. Williams & Wilkins Co., Baltimore, 1973.
6. Keyes, J. L. "Basic Mechanisms Involved in Acid-Base Homeostasis." *Heart and Lung* 5:239–246, 1976.
7. Keyes, J. L. "Blood-Gas Analysis and the Assessment of Acid-Base Status." *Heart and Lung* 5:247–255, 1976.
8. Pitts, R. F. *Physiology of the Kidney and Body Fluids*, 3rd ed. Yearbook Medical Publishers, Chicago, 1974 (Chapter 10).
9. Siggaard-Andersen, Ole. *The Acid-Base Status of the Blood*, 4th ed. Williams & Wilkins Co., Baltimore, 1974.
10. Swan, R. C. and R. F. Pitts. "Neutralization of Infused Acid by Nephrectomized Dogs." *Journal of Clinical Investigation* 34:205–212, 1955.

11. Swan, R. C., D. R. Axelrod, M. Seip, and R. F. Pitts. "Distribution of Sodium Infused into Nephrectomized Dogs." *Journal of Clinical Investigation* 34:1795, 1955.

CHAPTER 6
Pathophysiology of Acid-Base Disturbances

1. Cohen, J. J. and J. P. Kassirer. *Acid-Base*. Little, Brown, and Co., Boston, 1982.
2. Davenport, H. W. *The ABC of Acid-Base Chemistry*, 6th ed. University of Chicago Press, Chicago, 1974.
3. Greenberger, N. J. *Gastrointestinal Disorders*, 2nd ed. Yearbook Medical Publishers, Chicago, 1981.
4. Hills, A. G. *Acid-Base Balance*. Williams & Wilkins Co., Baltimore, 1973.
5. Kassirer, J. P. "Serious Acid-Base Disorders." *New England Journal of Medicine* 291:773–776, 1974.
6. Keyes, J. L. "Blood-Gas Analysis and the Assessment of Acid-Base Status." *Heart and Lung* 5:247–255, 1976.
7. Nahas, G. G. (ed.). "Current Concepts of Acid-Base Measurement." *Annals of New York Academy of Science* 133:1–274, 1966.
8. Siggaard-Andersen, Ole. *The Acid-Base Status of the Blood*, 4th ed. Williams & Wilkins Co., Baltimore, 1974.

CHAPTER 7
Pathophysiology of Disturbances in Regulation of Volume and Osmolality of Body Fluids

1. Pitts, R. F. *Physiology of the Kidney and Body Fluids*, 3rd ed. Yearbook Medical Publishers, Chicago, 1974.
2. Robbins, S. L., M. Angell, and V. Kumer. *Basic Pathology*. W. B. Saunders Co., Philadelphia, 1981 (Chapters 2 and 3).
3. Rose, B. D. *Clinical Physiology of Acid-Base and Electrolyte Disorders*. McGraw-Hill Book Co., New York, 1977.
4. Schrier, R. W. (ed.). *Renal and Electrolyte Disorders*, 2nd ed. Little, Brown, and Co., Boston, 1980.
5. Thier, S. O. "The Kidney," in L. H. Smith and S. O. Thier (eds.), *Pathophysiology: The Biological Principles of Disease*. W. B. Saunders Co., Philadelphia, 1981.
6. Valtin, H. *Renal Dysfunction*. Little, Brown, and Co., Boston, 1979.
7. Vander, A. J. *Renal Physiology*, 2nd ed. McGraw-Hill Book Co., New York, 1980.

CHAPTER 8
Electrolyte Balance and Imbalance

1. Berne, R. M. and M. N. Levy. *Cardiovascular Physiology*, 4th ed. C. V. Mosby Co., St. Louis, 1981.
2. Pitts, R. F. *Physiology of the Kidney and Body Fluids*, 3rd ed. Yearbook Medical Publishers, Chicago, 1974.
3. Popovtzer, M. M. and J. P. Knochel. "Disorders of Calcium, Vitamin D, and Parathyroid Hormone Activity," in R. W. Schrier (ed.), *Renal and Electrolyte Disorders*, 2nd ed. Little, Brown, and Co., Boston, 1980.

4. Rose, B. D. *Clinical Physiology of Acid-Base and Electrolyte Disorders*. McGraw-Hill Book Co., New York, 1977.
5. Thier, S. O. "The Kidney," in L. H. Smith and S. O. Thier (eds.), *Pathophysiology: The Biological Principles of Disease*. W. B. Saunders Co., Philadelphia, 1981.
6. Valtin, H. *Renal Dysfunction*. Little, Brown, and Co., Boston, 1979.
7. Vander, A. J. *Renal Physiology*, 2nd ed. McGraw-Hill Book Co., New York, 1980.

CHAPTER 9
Hypoxia and Venous Blood-Gas Composition

1. Carveth, M. L., A. J. Schriver, W. E. Peterson, and J. L. Keyes. "Acid-Base Assessment with Peripheral Venous Blood." *Heart and Lung* 13:48–54, 1984.
2. Comroe, J. H., Jr. *Physiology of Respiration*, 2nd ed. Yearbook Medical Publishers, Chicago, 1974.
3. Griffith, K. K., M. B. McKenzie, W. E. Peterson, and J. L. Keyes. "Mixed Venous Blood-Gas Composition in Experimentally Induced Acid-Base Disturbances." *Heart and Lung* 12:581–586, 1983.
4. Kazarian, K. K. and L. R. Del Guercia. "The Use of Mixed Venous Blood-Gas Determinations in Traumatic Shock." *Annals of Emergency Medicine* 9:179–182, 1980.
5. Kryger, M. H. *Pathophysiology of Respiration*. John Wiley & Sons, New York, 1981.
6. Michel, C. C., "The Buffering Behavior of Blood During Hypoxemia and Respiratory Exchange: Theory." *Respiration Physiology* 4:283–291, 1968.
7. Murphy, J., B. Berner, and J. L. Keyes. "Mixed Venous Blood-Gas Composition in Dogs During Hypovolemia," in preparation.
8. Murray, J. F. *The Normal Lung*. W. B. Saunders Co., Philadelphia, 1976.
9. Roos, A. and C. Thomas. "The in Vivo and in Vitro Carbon Dioxide Dissociation Curves of True Plasma." *Anesthesiology* 28:1048–1063, 1967.
10. Slonim, M. B. and L. H. Hamilton. *Respiratory Physiology*, 4th ed. C. V. Mosby Co., St. Louis, 1981.
11. Tung, S. H., J. Bettice, B. C. Wang, and E. B. Brown. "Intracellular and Extracellular Acid-Base Changes in Hemorrhagic Shock." *Respiration Physiology* 26:229–237, 1976.

CHAPTER 10
Clinical Examples

1. Cohen, J. J. and J. P. Kassirer. *Acid-Base*. Little, Brown, and Co., Boston, 1982.
2. Greenberger, N. J. *Gastrointestinal Disorders*, 2nd ed. Yearbook Medical Publishers, Chicago, 1981.
3. Kassirer, J. P. "Serious Acid-Base Disorders." *New England Journal of Medicine* 291:773–776, 1974.
4. Valtin, H. *Renal Dysfunction*. Little, Brown, and Co., Boston, 1979.

APPENDICES

1. Oh, M. S. and H. J. Carroll. "The Anion Gap." *New England Journal of Medicine* 297:814–817, 1977.

2. Pitts, R. F. *Physiology of the Kidneys and Body Fluids*. 3rd ed. Yearbook Medical Publishers, Chicago, 1974.
3. Siggaard-Andersen, Ole. *The Acid-Base Status of the Blood*, 4th ed. Williams & Wilkins Co., Baltimore, 1974.
4. Slonim, N. B. and L. H. Hamilton. *Respiratory Physiology*, 4th ed. C. V. Mosby Co., St. Louis, 1981.

Index*

**t* denotes table, *n* denotes footnote

[redacted]